Diamondoids

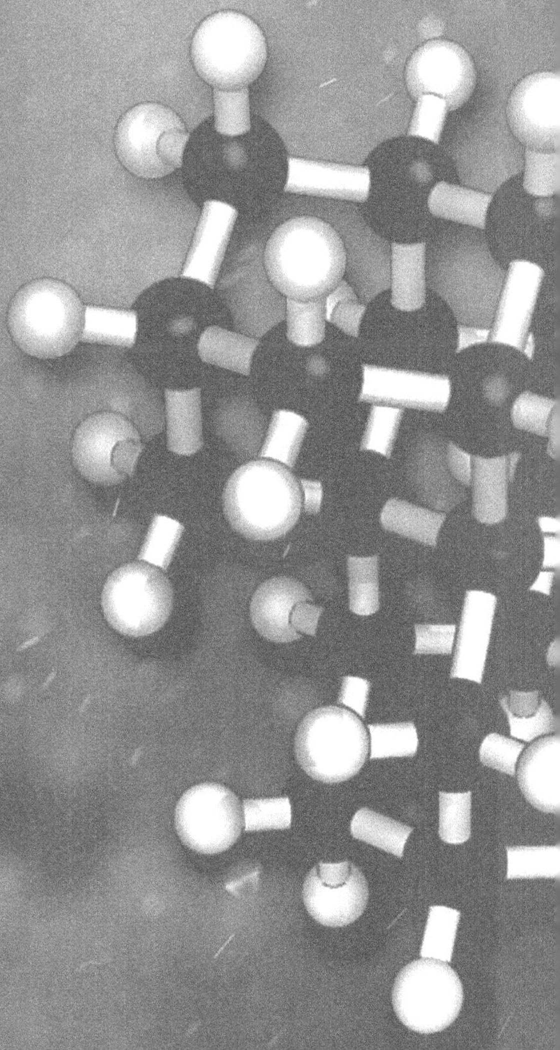

Diamondoids

Synthesis, Properties, and Applications

Sven Stauss
Kazuo Terashima

Pan Stanford Publishing

Published by

Pan Stanford Publishing Pte. Ltd.
Penthouse Level, Suntec Tower 3
8 Temasek Boulevard
Singapore 038988

Email: editorial@panstanford.com
Web: www.panstanford.com

British Library Cataloguing-in-Publication Data
A catalogue record for this book is available from the British Library.

Diamondoids: Synthesis, Properties, and Applications

ISBN 978-981-4745-18-5 (Hardcover)
ISBN 978-1-315-36466-7 (eBook)

MIX
Paper from responsible sources
FSC® C011748
www.fsc.org

Printed in the Great Britain by Ashford Colour Press Ltd

Contents

Part III Novel Approaches for the Synthesis of Diamondoids by Microplasmas

Symbols and Abbreviations

List of Symbols

C_p	Heat capacity	
E	Energy	
E_b	Binding energy	(eV)
E_{VBM}	Valence band maximum	
F_D	Density fluctuation	
M_r	Relative molecular weight	
p	Pressure	(Pa)
p_{crit}	Critical pressure	(Pa)
T	Temperature	(K)
T_{crit}	Critical temperature	(K)
V	Volume	(m^3)
V_{appl}	Applied voltage	(kV)
$V_{p\text{-}p}$	Peak-to-peak voltage	(kV)
β	Compressibility	
κ	Heat conductivity	

List of Abbreviations

CP	Critical point
B3LYP	Becke, three-parameter Lee–Yang–Parr exchange correlation functional used for DFT
DFT	Density functional theory
DOS	Density of electronic states
GC-MS	Gas chromatography–mass spectrometry

HOMO	Highest occupied molecular orbital
LUMO	Lowest unoccupied molecular orbital
SIM	Selected ion monitoring
SXE	Soft X-ray emission

Preface

Like a modern drama, one could consider this book to consist of three main acts: In the first act, we set the stage for the topic and expose the main actors, that is, diamondoids, and their relation to other carbon nanomaterials. In the same part, we delve more deeply into the chemical and physical properties of diamondoids and give an overview of their current and possible future applications.

In the second part of the book, we, little by little, approach the main complications related to the application of especially larger diamondoids: the current approaches for obtaining diamondoids and the current attempts to obtain them by conventional chemical synthesis. In particular, we aim at demonstrating the different problems associated with the various conventional approaches.

Finally, in the third part, we present possible alternative solutions for synthesizing diamondoids from the smallest member, adamantane, using plasmas generated in supercritical fluids. Because the field of plasmas generated in such high-density fluids is still not known to a wide audience, we are first briefly presenting the properties and applications of supercritical fluids before discussing the properties of electric discharge and pulsed laser plasmas generated in supercritical fluids.

The last part of the book deals with the application of such plasmas for diamondoids in more detail before giving a final assessment of the current state of the research on diamondoids.

Molecular diamonds, commonly called diamondoids, are a very interesting class of carbon nanomaterials that can be considered the archetypical molecular building block. While still not as widely known as other carbon nanomaterials, such as carbon nanotubes and graphene, previous and current research shows that this class of materials holds a lot of promise for many different fields of science and nanotechnology.

One of the goals in preparing this book was to introduce interested readers to the field of diamondoids; their structure, chemical, and physical properties; and the different isolation and synthesis approaches that currently exist. In addition to conventional isolation and synthesis approaches, we also introduce new ones, especially those based on electric discharge and pulsed laser plasmas generated inside supercritical fluids and at atmospheric pressure.

We hope that the present book will be useful not only as both an introductory and a reference text on diamondoids but also as an inspiration for further research on this remarkable class of nanomaterials.

This book is mainly aimed at researchers and graduate students who are curious about the field of diamondoids, that is, materials science, physics, and chemistry, but we hope that it will also be interesting for persons from other fields of science and technology.

One of our goals in writing this book was to make it an introductory text on diamondoids aimed at researchers and graduate students active in the field and at the same time to make it also a stand-alone text so that people who are not familiar with certain topics do not have to refer to the scientific literature themselves.

Special care was given to the figures and graphs. Whenever possible, we remade both graphs and figures in this book in order to avoid raster images as much as possible. In the cases where we have relied on data from other publications, we have indicated all the original data sources and references. We also tried to make the figure and table captions as self-explanatory as possible so that people who just want to flip through the book can understand the main points of an illustration or graph without having to resort to the main text.

For those readers who wish to delve more deeply into the field of diamondoids, we have compiled a comprehensive list of the current scientific literature. Especially the field of organic synthesis of diamondoids is very vast. However, for the convenience of the readers, we have limited ourselves to those references that can currently be accessed electronically.

In writing this book, we have tried to remain as general and make the contents as enduring as possible, but at the same time we also wanted to provide the reader with the latest results of the research on diamondoids. However, we are fully aware that with todays rapid pace with which science and technology advance, the contents of this book might become outdated quite rapidly. Nevertheless, we hope that this will not happen too soon and that it can serve researchers from many different backgrounds as a reference and hopefully also as an inspiration for their own research.

A very special thank-you goes to Stanford Chong for suggesting this book project and his patience and encouragement during the writing process. We are also indebted to all the staff at Pan Stanford Publishing for their efforts in bringing this book to print.

We would like to thank the following persons for their support in the work related to this book:

- Prof. Takehiko Sasaki for his help with mass spectrometry measurements and comments concerning the interpretation of the results
- Prof. Hiroshi Kataoka and Dr. Minoru Suzuki (now at Shimadzu Corp.) for their support with gas chromatography–mass spectrometry measurements and helpful discussions
- Prof. Tohru Suemoto and Prof. Motoyoshi Baba (now at Saitama University) for their help with pulsed laser ablation
- Dr. Keiichiro Urabe (now at Air Liquide) for commenting on one of the very early drafts of the manuscript
- All present and former students of the Terashima Lab for the work presented in this text

Concerning the experiments on diamondoid synthesis, we would like to thank Dr. Takaaki Tomai (now at Tohoku University), Dr. Hiroyuki Miyazoe (now at IBM Thomas J. Watson Research Center), Hirokazu Kikuchi, Sho Nakahara, Tomoki Shizuno, Fumita Oshima, Chikako Ishii, Satoshi Kato, and Koichi Kuribara. We would also like to thank Dr. David Z. Pai (Université de Poitiers) for his contribution to atmospheric-pressure diamondoid synthesis.

Since the topic of diamondoids encompasses various branches of science and readers are most probably from many different fields of science, we have tried to keep the text as self-explanatory as possible. Especially organic chemists might find our treatment of organic synthesis too basic; however, as some of the readers (and the authors) do not have a background in organic chemistry, we hope that they can bear with us.

As any book, probably this one also contains many errors and oversights, and any mistakes in the text are entirely the responsibility of the authors. We are grateful for any feedback from the readers, and we are grateful for any feedback from the readers to point us to these.

Finally, we hope that the book can serve both as a reference text and an inspiration, both for seasoned researchers and new students, and that the field of diamondoids can grow more in the future.

This work was supported financially in part by a Grant-in-Aid for Scientific Research on Innovative Areas (Frontier Science of Interactions between Plasmas and Nano-interfaces, Grant No. 21110002) from the Ministry of Education, Culture, Sports, Science, and Technology of Japan. The authors thank the Materials Design and Characterization Laboratory, Institute for Solid State Physics, University of Tokyo, for providing access to the pulsed laser facility.

Sven Stauss
Kazuo Terashima
Kashiwa and Tokyo, Japan

Part I

Diamondoids: Structures, Properties, and Applications

Chapter 1

Introduction

With the turn of the 21st century, the fabrication and synthesis of new materials have become paramount either for further enhancing existing or for enabling new applications. For example, the combination of new process technologies and the use of new semiconductor materials has allowed decreasing the characteristic size of functional components in electronic circuits and memories and to realize "more than Moore."

Since the demonstration of the first working transistor at Bell Laboratories by John Bardeen, Walter Brattain, and William Shockley in 1947, the number of transistors that can be packed on a chip has doubled roughly every 18 months. The relationship between this duration and the doubling of the number of transistors on a chip has become known as Moore's law (Moore, 1998). The evolution of the number of transistors as a function of time and in comparison to Moore's law is depicted in Fig. 1.1.

Up to now, with the advancement and optimization of new processing technologies, the characteristic sizes of transistors are approaching the order of a few nanometers. Currently, the characteristic dimensions of electronic components are reaching orders of a few nanometers, and while recent new applications and nonplanar transistor designs, for example, fin-shaped field-

Diamondoids: Synthesis, Properties, and Applications
Sven Strauss and Kazuo Terashima

ISBN 978-981-4745-18-5 (Hardcover), 978-1-315-36466-7 (eBook)
www.panstanford.com

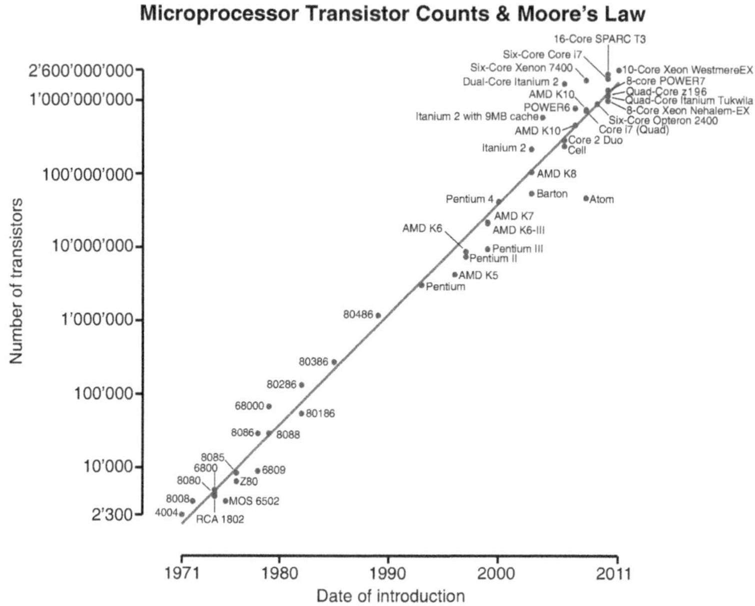

Figure 1.1 Evolution of the number of transistors. The graph displays the variation from 1970 to 2011, together with microprocessor families. The solid line corresponds to the count of transistors doubling every two years. Data adapted from Moore (1998).

effect transistors (finFETs), allow overcoming certain limitations of conventional two-dimensional silicon-based transistors, ultimately, the current approach will reach its physical limit that cannot be overcome and further shrinking of transistors will not be possible anymore.

The advancement and decrease of the feature sizes are driven by not only the optimization of the processing steps, a decrease of the wavelength in the light sources used for lithography, etc., but also the integration and combination of new materials. In contrast to conventional top-down approaches, where functional components are fabricated using macroscopic masks, etc., bottom-up processing has been suggested as an alternative. In this approach, functionality is achieved by using molecular building blocks consisting of nanoparticles, nanowires, etc.

Similar behavior has been observed also for other electronic components, for example, hard disks, where the capacity has been shown to double in a similar fashion, or the resolution of displays, where pixel densities are also increasing to the extent that individual pixels cannot be distinguished anymore by the naked eye.

The trend of this exponential increase in the number of transistors or other electronic components has been driven by improvements of processing, for example, the decrease of the characteristic feature sizes using light sources of smaller and smaller wavelengths.

Currently, characteristic dimension of electronic components reaching orders of a few nanometers, and while recent new applications, for example, finFETs, allow to overcome certain limitations of conventional transistors, ultimately, the current approach will hit a dead end.

Because of this, in recent years, other types of processing techniques, so-called bottom-up, have gained increasingly interest as an alternative to conventional top-down fabrication methods. Top-down techniques heavily rely on repetitive of patterning, deposition, and etching, the feature size being determined by the wavelength of the light source used during patterning, as well as characteristics of, for example, the plasma chemistry. In bottom-up techniques, larger building blocks are obtained by guided self-assembly of atoms, clusters, or wires (cf. Fig. 1.2.)

Currently, predictions of the shrinking of characteristic feature sizes in microchips have become less aggressive, concomitant with a reduction in the pace of new technologies. The main difficulty in further reducing gate widths is the limits imposed by conventional microfabrication processes. The physical limit of downsizing of gate widths is expected to be reached at 5 nm, because leakage currents become too high (Iwai, 2009).

On the other hand, it has been proposed that novel materials, for example, novel III–V compound semiconductors, could permit one to solve some of the problems encountered by conventional silicon-based devices (del Alamo, 2011).

Nanomaterials are gaining increasing interest and are finding more and more their way from scientific laboratories to industrial applications. Among these are Si nanowires, which have shown

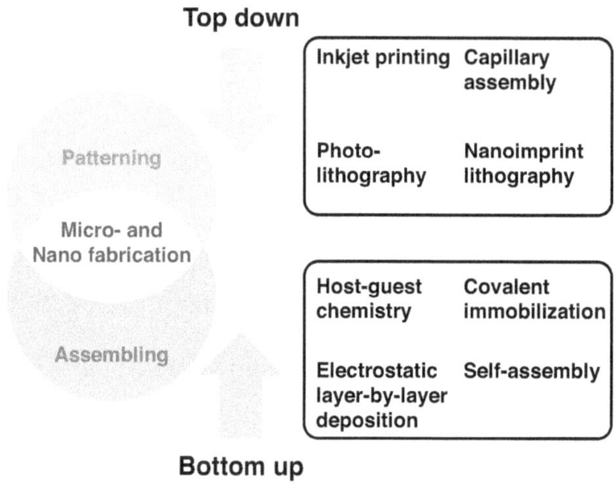

Figure 1.2 Illustration of conventional top-down and bottom-up processes. Figure adapted from Curri et al. (2010).

promise as material transistors, for sensors but also for enhancing the efficiency of photovoltaic cells. Because Si can store about 10 times more Li when compared to graphite, it also has shown promise for use in Li ion batteries.

Another very important field of research that has become more and more important is biotechnology. To develop new pharmaceuticals and also to understand molecular processes occurring inside cells, organs, and bodies of organisms, new methods have to be devised. For observing molecular processes inside cells, different types of nanomarkers are used, such as fluorescent nanoparticles. Such nanomarkers typically consist of nanoparticles that have been functionalized by attaching a fluorescent molecule to them or that already contain color centers that are fluorescent (e.g., diamonds that contain nitrogen vacancy defects). One constraint that such nanoparticles have to fulfill is that they are biocompatible and that they are nontoxic. Conventionally, functionalized inert metallic nanoparticles such as gold have been used as fluorescent particles.

In addition to metallic nanoparticles, semiconductor nanoparticles are gaining more and more attention due to the many different possibilities that are offered by their doping.

Besides these types of nanomaterials, those based on carbon are one of the most important. Carbon is one of the most abundant elements on earth, and owing to the possibility of different hybridization, many different nanomaterials can be obtained, such as fullerenes, carbon nanotubes, graphene, and nanodiamonds (the relation between these different classes will be discussed in detail in Chapter 2. In addition to these, in recent years nanodiamonds, with sizes in the range of ~5–20 nm have been used increasingly, owing to their nontoxicity and biocompatibility. In addition, the presence of carbon allows functionalization with relative ease. The main topic of this book is a special class of carbon nanomaterials, diamond molecules, also called diamondoid hydrocarbons or simply diamondoids.

In contrast to other types of nanomaterials, for example, metallic nanoparticles but also nanodiamonds, their structure and sizes are well defined, and they possess a very high thermal and structural stability.

The topic of this book is a special class of carbon nanomaterials, diamond molecules, synonymously called diamondoid hydrocarbons, polymantanes, or simply diamondoids. Diamondoids are aliphatic cage structures, that is, they consist of a carbon framework saturated completely by hydrogen terminations.

Because of their well-defined structure, they represent the archetypical molecular building block, and as we will show in subsequent parts of this book, this is one reason why they show promise for a wide range of applications where they could probably replace currently used nanomaterials.

The contents of this book are structured as follows: In Chapter 2, we introduce diamond molecules—diamondoids—and discuss their relation to other carbon nanomaterials (carbon nanotubes, graphenes, etc.). We also give a brief introduction to the nomenclature of diamondoids and briefly discuss the molecular symmetries of diamondoids and the crystalline structure of solid diamondoids. Chapter 3 properties of diamondoids.

In Chapter 4, we present current and possible future applications of diamondoids in various scientific and technological fields, medicine, pharmaceutics, and opto- and nanoelectronics. We also present a few topics where possibly one day diamondoids could replace currently used macroscale diamonds or nanodiamonds, namely in quantum computing and quantum communication.

In the second part of this book, that is, Chapters 5 and 6, we present the current approaches to isolating diamondoids from oil and gas reservoirs and the approaches to obtaining them by organic chemical synthesis.

In the final part of the Chapter 7 to 9, we present alternative approaches for the synthesis of diamondoids, which are based on plasmas generated in supercritical fluids and at atmospheric pressure.

We conclude the book with a brief summary of and an outlook on possible future directions in the field of diamondoids, their synthesis, and their applications.

Chapter 2

Structure, Nomenclature, and Symmetry of Diamondoids

In this chapter, we first introduce the different structures of carbon nanomaterials that arise as a result of their hybridization. A part of this chapter is also devoted to the classification and nomenclature of diamondoids, and the crystal structures of diamondoids. Finally, we also make a brief comparison between diamondoids and nanodiamonds, and point out their principal analogies and differences. As will be explained in more detail in this chapter, compared to the macroscale counterparts, that is, diamond, in nanodiamonds and diamondoids, both the size and surface chemistry impart electronic and magnetic properties (Enoki et al., 2009). A more general treatise on carbon nanomaterials and nanotechnology can be found in (Krueger, 2010).

2.1 Diamondoids and Their Relation to Other Carbon Nanomaterials

Estimates place carbon as the fourth-most abundant element in the sun and solar system (Atkins et al., 2010), and carbon at the

Diamondoids: Synthesis, Properties, and Applications
Sven Strauss and Kazuo Terashima

ISBN 978-981-4745-18-5 (Hardcover), 978-1-315-36466-7 (eBook)
www.panstanford.com

macroscopic scale can exist under different forms or allotropes: amorphous carbon, graphite, cubic diamond, and hexagonal diamond (Lonsdaleite).

In the last 50 years, other members of carbon allotropes at the nanoscale have been discovered: linear carbon chains such as carbyne/polyyne, fullerenes, carbon nanotubes, graphene, nanodiamonds and diamondoids. The reason for the ability of carbon to form different chemical bonds that result in different allotropes is related to the electronic structure of carbon. In carbon, with the atomic number 6, the occupation of the different orbitals in the ground state is $1s^2 2s^2 2p_x^1 2p_y^1$. The 2s and 2p valence orbitals of carbon can be combined to form different hybridized orbitals, leading to the formation of sp^1, sp^2, and sp^3 orbitals. Figure 2.1

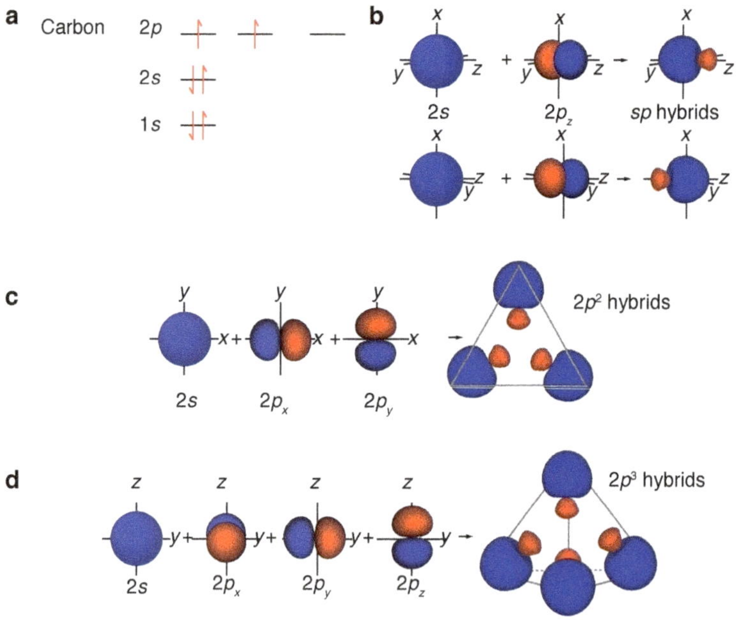

Figure 2.1 Electronic configuration and hybridization of carbon. (a) Occupation of core and valence shells in carbon. The linear combination of the 2s and 2p orbitals leads to the formation of different orbitals. (b) Combination of one 2s and one 2p orbital: Formation of sp^1 orbital. (c) Formation of sp^2 orbital by the combination of one 2s and two 2p orbitals. (d) Combination of one 2s and $2p_x$, $2p_y$, and $2p_z$ orbitals leads to the formation of sp^3, whose principal axes point to the edges of a tetrahedron.

illustrates the electron occupation of the orbitals of C atoms and the formation of sp^1-, sp^2-, or sp^3-hybridized orbitals. In the case of the sp^3 orbitals, the lobes point to the edges of a tetrahedron and are separated by an angle of 109.5°.

The classification of carbon materials based on the dimensionality of their structure, and their basic properties, has been discussed in an extensive review (Shenderova et al., 2002). Figure 2.2 illustrates the different carbon nanostructures that result from

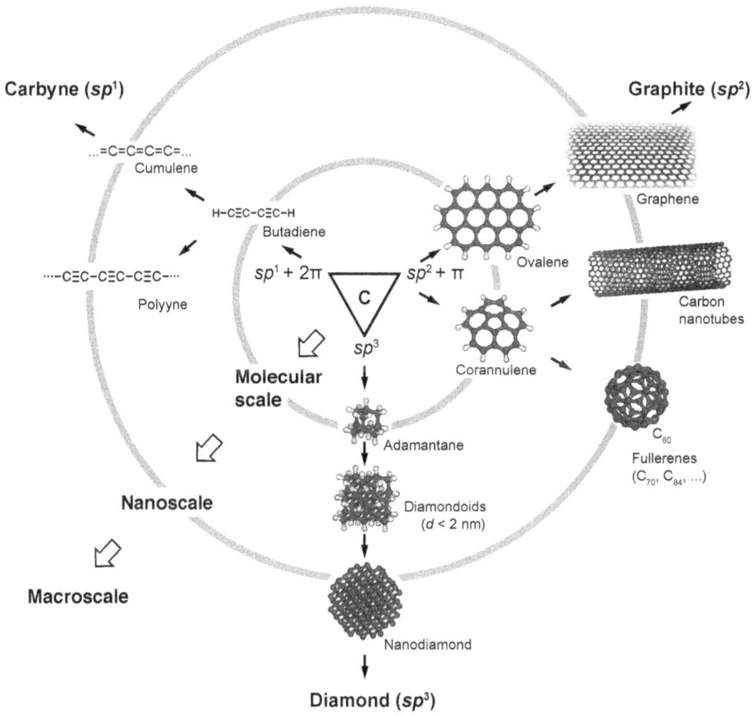

Figure 2.2 Overview of carbon nanostructures formed as a consequence of the different hybridizations of C, sp^1, sp^2, and sp^3. For sp^3-hybridized carbon nanostructures, corannulene can be considered as a building block for fullerene and carbon nanotubes, whereas ovalene can be considered as a graphene precursor. In contrast to fullerenes, carbon nanotubes, and graphene, the C–C bonds in diamondoids (defined to have characteristic sizes $d < 2$ nm) are sp^3 hybridized and the building block can be considered to be adamantane. Figure adapted from Shenderova et al. (2002).

the different hybridizations of carbon. As shown in the figure, C(sp^1)–C(sp^1) bonds lead to linear structures such as cumulenes and polyynes. The former consist of at least three consecutive C double bonds, whereas in the latter, the linear chain is formed by alternating single and triple C bonds.

For C(sp^2)–C(sp^2) bonds, a large variety of planar or curved structures can be formed—fullerenes, carbon nanotubes (CNTs), or graphene.

In contrast to these well-known carbon nanomaterials, especially fullerenes, CNTs and graphene, the structure of macroscopic diamond, nanodiamonds, and molecular diamond—also called diamondoid hydrocarbons or simply diamondoids—consists of C(sp^3)–C(sp^3)-hybridized bonds. The structure of diamondoids are formed by carbon cages that can be superimposed on a diamond lattice. As will be explained in more detail in the following sections, in contrast to nanodiamonds, the terminations of the carbon cage structure are well-defined and usually consist of hydrogen atoms.

2.2 The Structure of Diamondoids

The first diamondoid member is called adamantane, after the Greek word for diamond ($\alpha\delta\acute{\alpha}\mu\alpha\varsigma$—*adámas*, "unconquerable," "indestructible"). Adamantane has the chemical formula $C_{10}H_{16}$ and consists of a single cage that can be formed by fusing cyclohexanes in chair formation. All the C atoms in the structure of adamantane are C(sp^3)–C(sp^3)-hybridized, the dangling bonds being terminated by H atoms.

The relation between the structure of adamantane and the diamond lattice is illustrated in Fig. 2.3a, where the carbon framework of one adamantane is superimposed on the one of the diamond. As can be seen in Fig. 2.3b, which displays adamantane as both stick and van der Waals sphere models, it is highly symmetric, its point group being T_d (the symmetry elements belonging to this group and the implications for the interpretation of vibrational spectra will be discussed in Sections 2.5 and 3.3.1). As we will see in a following chapter, this high symmetry results in physical

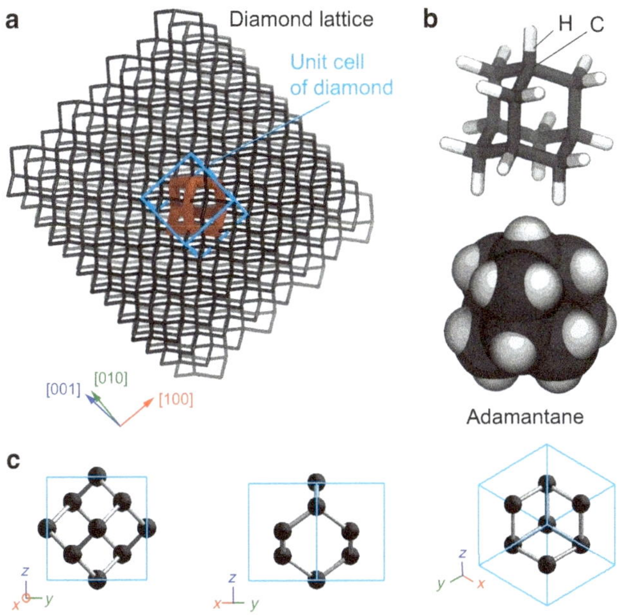

Figure 2.3 Relation of cubic diamond lattice and adamantane. (a) The unit cell and a 577-carbon-atom lattice of cubic diamond along the main crystallographic directions [100], [110], and [111] and its relation to the molecular structure of the smallest diamondoid, adamantane, whose carbon framework is indicated in red. (b) Adamantane, displayed in both stick and van der Waals sphere representations, consists of 10 carbon and 16 hydrogen atoms and is the basic unit of diamondoid hydrocarbons. (c) Adamantane in relation to diamond unit cell along three different viewing directions [100], [110], and [111]. For clarity, the hydrogen atoms are not displayed.

properties that set adamantane apart from hydrocarbons of similar molecular mass. Finally, the unit cell of the diamond accommodates adamantane, whose apical atoms occupy the centers of faces of the diamond unit cell (cf. Fig. 2.3c).

As will be explained in the next section, the distinction between diamondoids and nondiamondoids is made based on whether a molecule consists of at least one complete adamantane unit. We will discuss in more detail the difference between diamondoids and nondiamondoids and then explain the classification of diamondoids consisting of face-fused adamantanes, so-called polymantanes, which are the main topic of this book.

2.3 Classification of Diamondoids

In the discussion of the classification of diamondoids, we follow the treatment that has been presented in the seminal paper on diamondoids by Balaban and von Schleyer (Balaban and Schleyer, 1978).

Diamondoids or diamond hydrocarbons are generally defined as "molecules whose arrangement of C atoms allows total or partial superposition on the diamond lattice." However, this is true for many hydrocarbons that are not typically associated with diamondoids. Molecules whose structures can be superimposed on the diamond lattice include, for example, cyclohexane, decalin, simple alkanes (including methane), and also structures close to the diamond lattice such as cyclodecane, and bicyclo[3.3.1]nonane, but since they do not include at least one complete adamantane unit, they are considered as nondiamondoids.

In contrast, structures like the adamantane dimer belong to the class of diamondoids. Diamondoids are therefore defined as "hydrocarbons possessing at least one adamantane unit which are totally or largely superimposable on the diamond lattice" (Balaban and Schleyer, 1978). The distinction between diamondoid and nondiamondoid hydrocarbons and a few representative examples is illustrated in Fig. 2.4. For the sake of clarity, in the representations of these structures, the hydrogen terminations are not displayed. As shown in the figure, only molecules that consist of at least one complete adamantane unit, are classified as diamondoid hydrocarbons.

Next, we will examine more in detail the different types of molecular structure that are assigned to diamondoid hydrocarbons. The flowchart in Fig. 2.5 illustrates the separation into different subgroups, depending on whether the adamantane containing molecular structures are only partially or completely superimposable on the diamond lattice, and how many elements between adamantane units—atoms, edges, and faces—are shared.

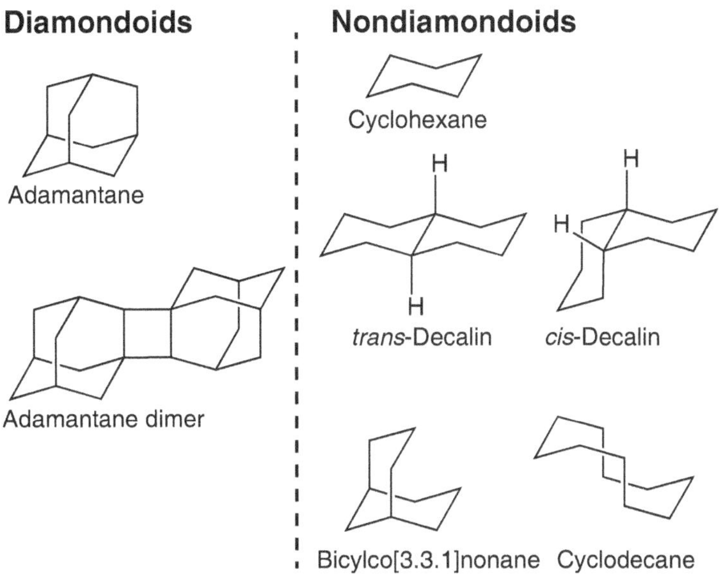

Figure 2.4 Distinction between diamondoid and nondiamondoid hydrocarbons. Diamondoids are hydrocarbons consisting of at least one complete adamantane unit. In contrast, structures that can also be superimposed on the diamond lattice but that do not contain one complete adamantane unit do not belong to the class of diamondoids. After Balaban and Schleyer (1978).

Molecules that contain at least one adamantane unit can be classified into the two main groups I and II:

I: those that are partially superimposable
II: those that are completely superimposable

on the diamond lattice. For molecules belonging to group II, they are further classified into two subgroups, depending on whether

(1) Not all atoms of the molecule belong to adamantane units
(2) All atoms are members of an adamantane unit

Members of group (2) can be further divided into those that consist of either

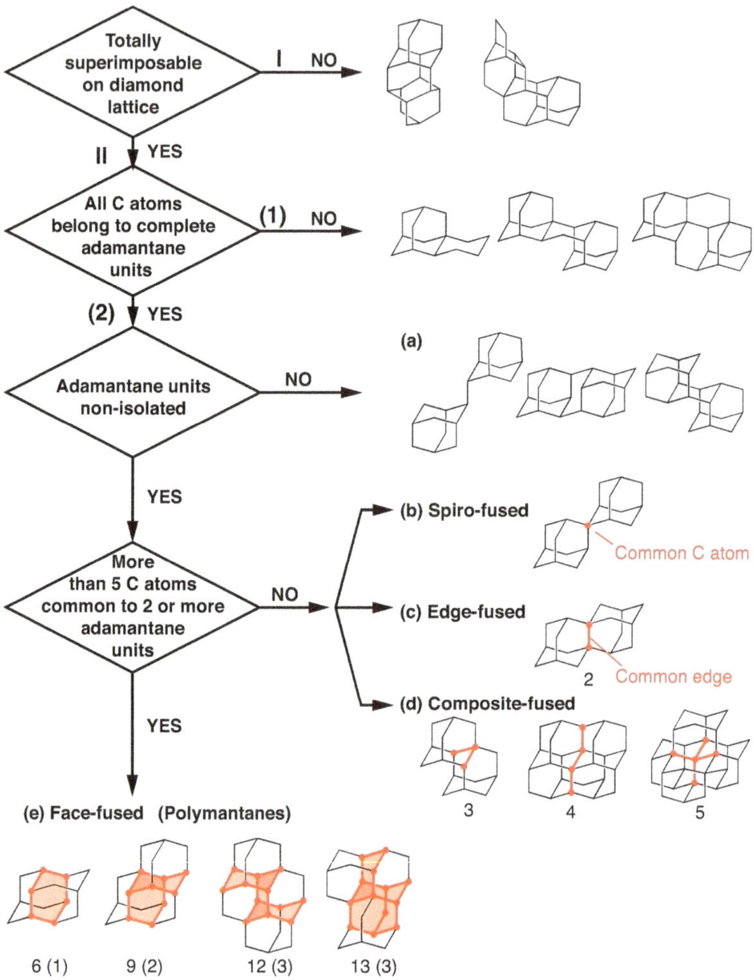

Figure 2.5 Flowchart illustrating the classification of diamondoid hydrocarbons or diamondoids. The classification here follows the exposition in Balaban and Schleyer (1978). The two main groups are those that are only partially superimposable on the diamond lattice (group I) and those that possess molecular structures that can be completely superimposed (group II). With increasing number of shared C atoms and edges, one finally obtains structures where adamantane units share faces, and these types of structures are named polymantanes. In the case of groups (b–d), the numbers indicate the C atoms shared, whereas in (e), the numbers indicate the shared C atoms and the number of shared faces (in parentheses).

(a) Nonisolated adamantane units

and those that are either fused by

(b) One common C atom (spiro fused)
(c) Two common C atoms (fused by one edge of the adamantane cage)
(d) Either 3, 4 or 5 common C atoms (2, 3 or 4 shared edges)

Finally, the last group is

(e) Two adamantane have 6 common C atoms, that is, sharing one face of the adamantane unit

These face-fused structures are called **polymantanes** and they are the class of diamondoid hydrocarbons that are the main topic of this text. The molecular structures of groups (b–e) displayed in Fig. 2.5 illustrate the increasing number of shared C atoms and edges of the adamantane cage (the numbers below the molecules indicating the number of shared C atoms and number of shared edges).

Following the smallest of the polymantanes, adamantane (cage number $n = 1$), the next larger member is diamantane that consists of two face-fused adamantane units ($n = 2$). As will be explained in more detail, the commonly accepted nomenclature of polymantanes uses Greek numerals as prefixes, followed by -mantane, that indicate the number of fused adamantane units. By face-fusing three adamantanes, one obtaines triamantane ($n = 3$). Starting with tetramantane ($n = 4$), structural isomers can be formed and the possibilities of arranging the diamondoid cages increases. Consequently, diamondoids with $n < 4$ are also called lower diamondoids, whereas for $n \geq 4$, they are called higher diamondoids. The details of the nomenclature and detailed classification of diamondoids will be discussed below.

Looking at the structures in Fig. 2.6, it becomes evident that for an increasing number of cages n, that the possibilities of the arrangement of adamantane units also increases. In this figure, the lines and vertices inside the structures of the diamondoids, represent the dualist graphs, which will be treated in detail in Section 2.4.2.

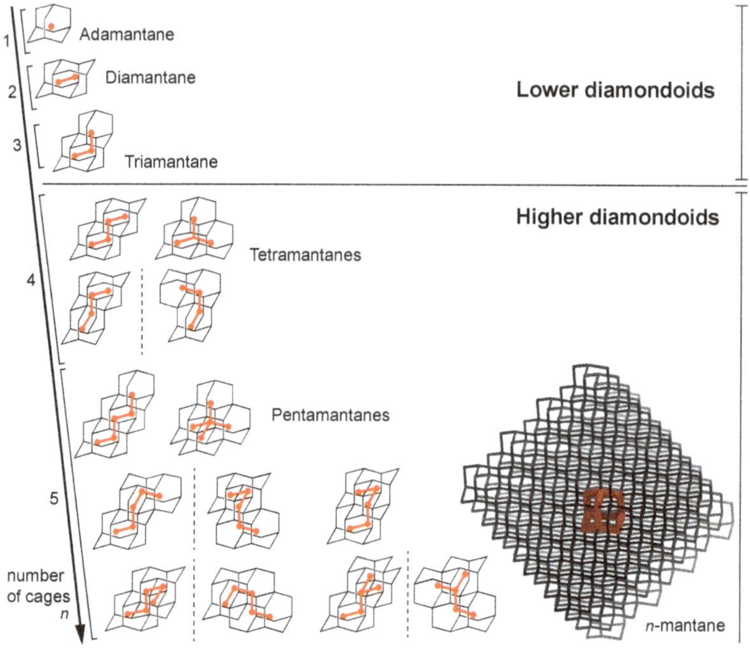

Figure 2.6 Molecular structures of lower and higher diamondoids and their relation to the diamond lattice. Structures of the lower diamondoids adamantane, diamantane, triamantane and their relation to an idealized 577-carbon diamond lattice (it is supposed that the diamond is terminated by hydrogen atoms, which are not shown for the sake of clarity). For the diamondoids from $n = 1$ to $n = 5$ the dualist graphs used in the systematic classification of diamondoids (Balaban and Schleyer, 1978) are also indicated as lines connecting the vertices at the center of the adamantane unit. Diamondoids with $n < 4$ are named lower diamondoids, whereas for $n \geq 4$, they are called higher diamondoids.

2.4 Nomenclature and Classification of Diamondoids

In this section, we briefly introduce the two main naming conventions that have been adopted for diamondoids. The first part focuses on the systematic IUPAC naming scheme introduced by von Baeyer (Baeyer, 1900) for polycyclic compounds, while the second part introduces the naming scheme of polymantanes commonly used for diamondoids that was devised by Balaban and von Schleyer (Balaban and Schleyer, 1978), and which is based on dualist graphs.

2.4.1 The von Baeyer Naming Scheme

For illustrating the von Baeyer naming scheme, we briefly demonstrate its application for the first two diamondoid members, adamantane and diamantane. The official IUPAC name consists of three parts: a prefix, which indicates the number of closed rings in the molecule, a middle part that lists the lengths of carbon chains or bridges connected to the main ring, and finally a suffix which indicates the total number of carbon atoms.

Therefore, the procedure for assigning the name can be broken down into three distinct steps, (1), (2), and (3). In the first step, the number of closed cycles the molecule is made of has to be deduced. As shown in Fig. 2.7, adamantane consists of three closed cycles, with atom sequences $1-2-3-10-7-8$, $3-4-5-6-7-10$, and $1-8-7-6-5-9$ (cf. Fig. 2.7). Alternatively, the number of cycles can also be determined from the total number of faces and subtracting one, or the number of scissions that are necessary to make the molecule linear (Filik, 2010). For example, adamantane has four faces, and three scissions are necessary to make it linear. The Greek number of cycles then gives the prefix of the name, that is, *tricyclo-* in the case of adamantane.

The next step consists in determining the longest closed ring of the C framework in the molecule. For adamantane, the longest closed ring consists of 8 atoms (cf. step (2) for adamantane in Fig. 2.7). Then, the two main bridgeheads, which are those C atoms that are linked to three or four other carbons in the diamondoid, have to be determined. Because of its symmetry, adamantane has four bridgeheads at positions 1, 3, 5, and 7. From all possible choices, the two main bridgeheads are selected in a fashion that allows to obtain the longest bridges or chains of C atoms. Consequently, in the case of adamantane, the two main bridgeheads are at positions 1 and 5. Then the bridges and the number of carbon atoms in all the bridges joining the two main bridgeheads are counted. There are 3 bridges connecting the two main bridgeheads, two containing three

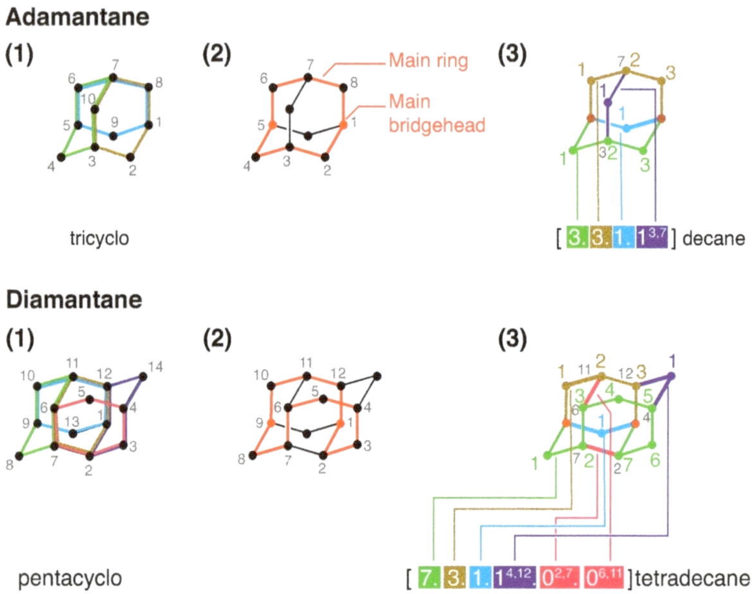

Figure 2.7 Example of the von Baeyer nomenclature of adamantane and diamantane. The procedure for finding the name consists basically of three steps: (1) Determine the number of closed cycles the molecule is made of; the Greek digit of the number of cycles is the prefix, for example, *tricyclo-* for three; (2) find the longest closed cycle of the molecule; assign the main bridgeheads (C atoms that are connected to at last three other C atoms); and (3) list the number of C atoms in the different bridges connected to the main bridgeheads. For C atoms that are not connected to the main bridgeheads, indicate the C atoms' positions that are at the start and end of the bridge as superscripts. Finally the last component of the name is the Latin digit of total C atoms in the molecule (e.g., *-decane* for adamantane and *-tetradecane* for diamantane).

atoms, and one bridge with one. There is one more bridge containing 1 C atom between positions 3 and 7, which is not connected to the main bridgeheads. In this case, the positions of the C atoms are added as superscripts and separated by commas to the number of C atoms. The numbers of bridge atoms are then placed in square brackets, separated by points: $[3.3.1.1^{3,7}]$. Finally, the Latin term for the total number of C atoms in the molecule is added as a suffix, that is, *-decane* for adamantane. Combining these three elements, the IUPAC name of adamantane is tricyclo[3.3.1.1^{3,7}]decane.

In the case of diamantane, the procedure is identical, except that identifying the largest closed cycle and the two main bridgeheads becomes more complicated, as illustrated in Fig. 2.7. As the molecular structure of diamantane contains five closed rings, its prefix is *pentacyclo-*. The longest cycle in the structure consists of 12 C atoms, and there are 8 bridgeheads (at positions 1, 2, 4, 6, 7, 9, 11, and 12). The two main bridgeheads are assigned to C atoms 1 and 9, and there are three bridges that are connected to them, one containing 7 C atoms (sequence $2-3-4-5-6-7$), one with three ($10-11-12$), and one with a single C atom (connecting C atoms 1 and 9). Then there is one bridge with one C atom linking atoms 4 and 12, and two bridges containing no C atom but linking C atoms 2 two 7 and 6 to 11. Finally, the von Baeyer designation of diamantane is pentacyclo[7.3.1.1^{4,12}.0^{2,7}.0^{6,11}]tetradecane.

The example of diamantane shows that for more complex diamondoid structures, the procedure becomes too involved to determine the corresponding von Baeyer name. There are a few methods that can simplify the procedure outlined above. One is based on using planar graphs, that is, the projection of the 3D molecular structure of a diamondoid along a given direction onto a plane (so-called Schlegel diagrams). This aids in determining the number of closed rings, the main ring in the structure, and the bridgeheads (Eckroth, 1967). As an example, Fig. 2.8 shows the structures and corresponding Schlegel diagrams for the first three diamondoids, adamantane, diamantane, and triamantane.

However, even these methods become too laborious for larger diamondoids. For example, the next diamondoid after diamantane, triamantane, is heptacyclo[7.7.1.1^{3,15}.0^{1,12}.0^{2,7}.0^{4,13}.0^{6,11}] octadecane. Consequently, for the naming of more complex diamondoids, computer algorithms are used (Balaban and Rucker, 2013), which has permitted to also correctly determine IUPAC names of functionalized diamondoids.

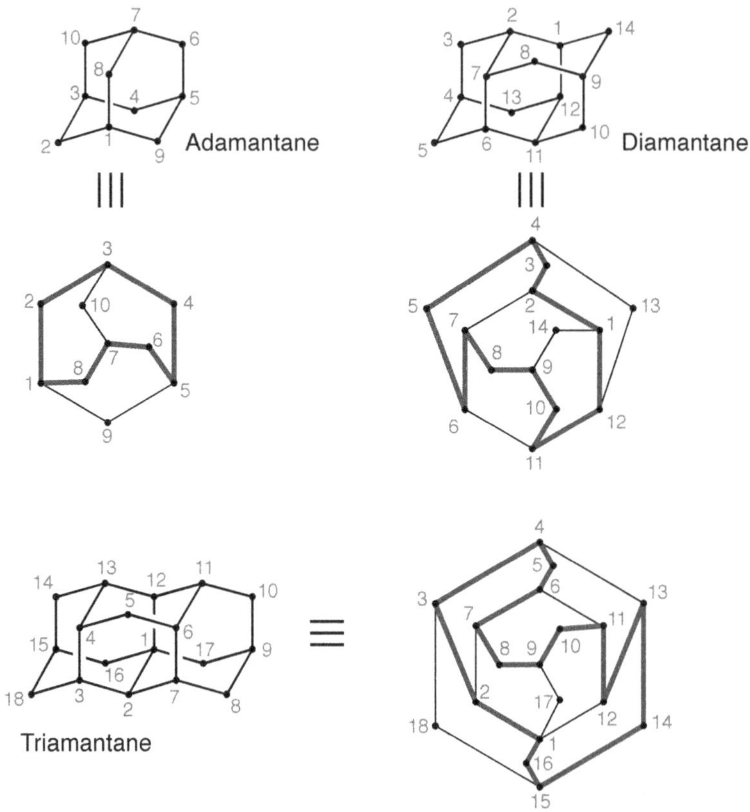

Figure 2.8 Molecular structures and corresponding Schlegel diagrams of adamantane, diamantane, and triamantane. Projecting the 3D structure along one axis on a plane allows one to identify the closed cycles, the longest ring and the main bridgeheads in the diamondoid structure more easily. After Eckroth (1967).

2.4.2 Nomenclature of Diamondoids Based on Dualist Graphs

While very flexible, for diamondoids, the von Baeyer scheme results in very unwieldy names, for example, pentacyclo[7.3.1.1^{4,12}.0^{2,7}.0^{6,11}]tetradecane for diamantane, and heptacyclo[7.7.1.1^{3,15}.0^{1,12}.0^{2,7}.0^{4,13}.0^{6,11}]octadecane for triamantane (Balaban, 2012). For this reason and to avoid the proliferation of trivial names that could have otherwise been introduced as alternatives to the von Baeyer naming scheme, Balaban and von Schleyer (Balaban and Schleyer, 1978) proposed an elegant and simple nomenclature based on dualist graphs. From the dualist graph codes, the structure of the diamondoids can be deduced much more easily.[a]

A dualist graph of a diamondoid can be likened to a linear, branched, or closed alkane, that itself can again be superimposed on the diamond lattice. In other words, a dualist graph of diamondoids is self-dualist. To generate a dualist graph, one starts from the center of an adamantane cage. One then connects this first vertex to the center of the next adjacent adamantane unit, following one of possible four directions (1, 2, 3, or 4), pointing to the edges of a tetrahedron (cf. Fig. 2.9). Adamantane has four faces, and because only these four directions are possible when connecting an adjacent adamantane unit, it is not necessary to indicate positive or negative directions, the direction becoming obvious from the position of the adamantane unit.

The procedure of connecting the vertices of the different adamantane units is then repeated until the whole diamondoid can be traced, adding the digits corresponding to the different directions to the dualist code. In the case of branched diamondoids, one chooses the permutation of the dualist code that represents the longest chain connecting the adamantane units of the diamondoid.

The number of chains linking the adamantane cages are placed in brackets, for example, [12] for triamantane, that is, the number of elements in the dualist code of an n-mantane consist of $n - 1$ digits.

When the diamondoid contains branches, the direction vectors pointing to these units from the main chain are placed into parentheses, for example, [1(2)3]tetramantane. When at a given vertex, two branches are present, they are placed in parentheses and separated by a comma, for example, [1(2, 3)4]pentamantane. In the case of longer branches, the chains of the branch are placed

[a] While dual graphs are frequently used in applied mathematics to describe the connectivity of closed surfaces, dualist graphs are different from dual graphs, since in dualist graphs, the distances and angles are fixed.

inside parentheses, without separating them by a comma, such as in [12(12)34]heptamantane.

Since there are always several permutations possible, a few conventions have been adopted. One is that the chosen sequence of direction vectors of the dualist graph are starting from the lowest indices. In addition, if the diamondoid contains branches, the elements in parentheses indicating these branches are moved as far to the end of the sequence as possible.

Figure 2.9 shows the structures and the construction of the dualist graphs for the lower diamondoids adamantane, diamantane, and triamantane, and all of the possible structures and corresponding dualist graphs of the first higher diamondoid, tetramantane.

Finally, [121(2)34]- and [12(12)34]heptamantane illustrate the convention of dualist graphs if a branch consists of more than one adamantane unit.

The list below summarizes additional rules for constructing a dualist graph of diamondoids (Balaban and Schleyer, 1978). Specifically, the following combinations are not allowed:

- . . . 111 . . .: Repeating digits
- . . . 1(1 . . .: Repeating digits
- . . . 1(2)1 . . .: Repeating digits before and after parentheses
- . . . 1(2, 3)1 . . .: Repeating digits before and after parentheses
- . . . 1(2, 3)2 . . .: Repeating two inside and after parentheses; only 4 is allowed in this instance
- . . . 1): The last digit inside a parenthesis cannot be 1
- 124 . . .: Lowest number rule (the last digit should be 3)
- 1214 . . .: Lowest number rule (the last digit should be 3)

As has been shown in the previous section, diamondoids can conceptually be formed by fusing together the basic adamantane units. Because the lower diamondoids (adamantane, diamantane, and triamantane) do not possess any structural isomers, only one permutation of the dualist graph is possible and consequently, the dualist codes of these three diamondoids are usually not written. From tetramantane ($n = 4$), the cages can be fused together by three different possibilities. As shown in Fig. 2.6, with increasing cage number n, the possibilities of cage arrangements increase. Furthermore, from diamondoids with $n \geq 6$, it is possible to

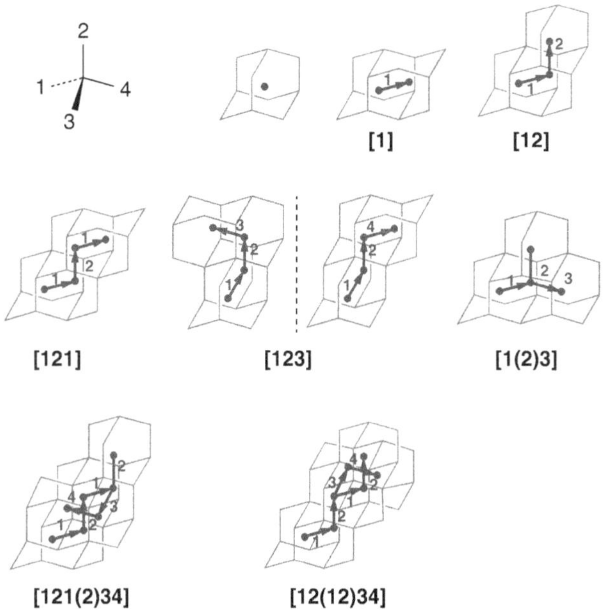

Figure 2.9 Construction of dualist graphs for polymantanes. The vertices of the dualist graph are the centers of the adamantane cages. A dualist graph of a diamondoid is formed by connecting the vertices of the adamantane centers, following four unit vectors pointing in the directions of a tetrahedron. Since for the lower diamondoids adamantane, diamantane, and triamantane, the arrangement of their cages is evident, usually their dualist codes are not used. In branched cases, the codes for the branch are placed in parentheses, as indicated for [121(2)34] and [12(12)34]heptamantane.

assemble adamantane units in such a fashion that their dualist graphs form cycles.

In the following, we will therefore briefly introduce the different types of polymantanes proposed (Balaban and Schleyer, 1978), which are based on whether their dualist graphs are acyclic (i.e., linear or containing branches), or whether they contain cycles.

In the case of polymantanes whose dualist graphs are acyclic, the authors (Balaban and Schleyer, 1978) have proposed the term "catamantanes" and "perimantanes" for molecules that contain cycles. For the latter, depending on whether the cycle in the dualist graph is composed of more than six vertices, the authors

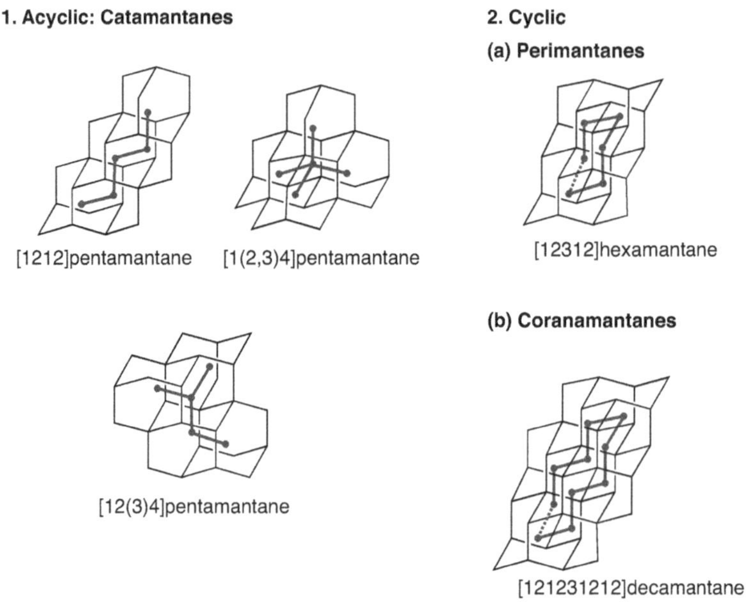

Figure 2.10 Classification of polymantanes according to their dualist graphs. The two main groups are those whose dualist graph is acyclic and those that contain cycles. If the vertices in a cycle of the dualist graph contain more than six C atoms, the molecule is attributed to the class of coronamantanes. The classification and examples are taken from Balaban and Schleyer (1978). The dashed line indicates the link between the first and the last cage; in the notation of the dualist graph, this is not written explicitly since it becomes evident from the sequence of the dualist graph notation.

have introduced the designation **coronamantane**. Examples of catamantanes, perimantanes and coronamantanes are displayed in Fig. 2.10. Just as a note, so far no coronamantanes have been described experimentally (Balaban, 2012).

For the sake of completeness, we summarize in Table 2.1 the dualist graph codes, point group symbols, chemical formulae, and total number of structures of polymantanes for $n = 1$ to 6. In this table, enantiomerism is not taken into account. As can be seen from Table 2.1, all the lower diamondoids ($n = 1$ to 3) possess only one structural isomer. Triamantane has 3 (4, if one

Table 2.1 List of polymantanes for cage numbers $n = 1$ to 6 (adamantane to hexamantane), their dualist codes, point group symbol, chemical formula, and total number. C: dualist code; S: point group symbol; F: chemical formula; NP: number of polymantanes, not taking into account enantiomers (Balaban and Schleyer, 1978). In the case of the regular and irregular catamantanes, the indicated chemical formulae correspond to both linear and branched diamondoids.

	Catamantanes										Perimantanes			
	Regular					Irregular					Irregular			
	Linear			Branched		Linear			Branched		Cyclic			
n	C	S	F	C	S	C	S	F	C	S	C	S	F	NP
1		T_d	$C_{10}H_{16}$											1
2	1	D_{3d}	$C_{14}H_{20}$											1
3	12	C_{2v}	$C_{18}H_{24}$											1
4	121	C_{2h}	$C_{22}H_{28}$	1(2)3	C_{3v}									3
	123	C_2												
5	1212	C_{2v}	$C_{26}H_{32}$	12(1)3	C_1									7
	1213	C_1		12(3)4	T_d	1231	C_s	$C_{25}H_{30}$						
	1234	C_2												
6	12121	C_{2h}	$C_{30}H_{36}$	121(2)3	C_1	12132	C_1	$C_{29}H_{34}$	12(1)31	C_s	12312	D_{3d}	$C_{26}H_{30}$	24
	12123	C_1		12(1)32	C_1									
	12131	C_1		121(3)4	C_s									
	12134	C_1		12(1)34	C_1	12314	C_1		123(1)2	C_1				
	12321	C_1		12(1,3)4	C_1				123(1)4	C_1				
	12324	C_2		12(3)12	C_s				12(3)41	C_s				
	12341	C_2		1(2)3(1)2	C_{2h}									
				1(2)314	C_1									
				12(3)14	C_1									
				1(2)3(1)4	C_2									

Table 2.2 List of heptamantanes ($n = 7$) with their dualist codes, point group symbol, chemical formula and total number. C: Dualist code; S: Point group symbol; F: Chemical formula; NP: Number of polymantanes, not taking into account enantiomers (Balaban and Schleyer, 1978)

	Catamantanes										Perimantanes			
	Regular, $C_{34}H_{40}$				Irregular						Irregular			
	Linear		Branched		Linear			Branched			Cyclic			
n	C	S	C	S	C	S	F	C	S	F	C	S	F	NP
7	121212	C_{2v}	1212(1)3	C_1	121231	C_1	$C_{33}H_{38}$	12(12)31	C_1	$C_{33}H_{38}$	121321	C_s	$C_{30}H_{34}$	88
	121213	C_1	121(2)31	C_1	121323	C_s		121(2)32	C_1		123124	C_s		
	121232	C_1	1212(3)4	C_s	121324	C_1		1213(1)2	C_1					
	121234	C_1	121(2)34	C_1	121341	C_1		12(1)3(1)2	C_1					
	121312	C_1	12(12)34	C_s	123132	C_1		12(1)314	C_1					
	121314	C_1	121(2,3)4	C_s	123134	C_1		12(1)3(1)4	C_1					
	121342	C_1	1213(1)4	C_1	123142	C_1		12(1)321	C_1					
	121343	C_2	121(3)21	C_1	123143	C_2	$C_{32}H_{36}$	1213(2)4	C_1					
	123214	C_1	121(3)23	C_1	123421	C_s	$C_{33}H_{38}$	12(1)341	C_1					
	123231	C_s	121(3)24	C_1				12(13)41	C_s					
	123234	C_2	12(1)324	C_1				12(1,3)41	C_s					
	123241	C_1	12(1)3(2)4	C_1				121(3)42	C_1					
	123412	C_2	12(13)32	C_3				12(1)342	C_1					
			12(13)34	C_1				12(13)43	C_1					
			121(3)41	C_1				123(1)23	C_1					

12(1,3)42	C_1	1(2)31(2)3	C_s	
121(3)43	C_1	123(1)24	C_1	
12(1,3)43	C_2	1(2)3124	C_1	
12(3)1(2)3	C_1	123(1,2)4	C_1	
12(3)124	C_1	1(2)31(2)4	C_1	
12(3)1(2)4	C_1	12(3)132	C_1	
1(2)3(1)24	C_1	1231(3)4	C_1	
1(2)3(1,2)4	C_s	12(31)41	C_{3v}	$C_{32}H_{36}$
1(2)3132	C_1	123(1)42	C_1	
12(3)134	C_1	12(3)142	C_1	$C_{33}H_{38}$
1(2)3134	C_1	1(2)3(1)42	C_1	
12(3)1(3)4	C_1	123(1)43	C_1	
1(2)31(3)4	C_2	123(2)14	C_1	
1(2)3142	C_1	1234(1)2	C_1	
1232(1)4	C_1	12(3)4(1)2	C_1	
123(2)41	C_1	1(2)34(1)2	C_{2v}	$C_{32}H_{36}$
12(3,4)12	C_{2v}			
123(4)21	C_1			

takes also into account chirality of the [123]tetramantane). In the case of pentamantane, ($n = 5$), one has 7 possible structural isomers, hexamantane ($n = 6$) has 24, and heptamantane ($n = 7$) has 88 (Balaban and Schleyer, 1978).

From pentamantane onwards, one can also see that it is possible to form diamondoids that do not possess the same molecular mass. Table 2.1 also lists the point groups the diamondoids belong to, and they will be discussed more in detail in Section 2.5.

2.4.3 Regular and Irregular Polymantanes

From the treatment in the previous sections on the structure of the diamondoids and their nomenclature using dualist graphs, one can see that the combinations of arranging the adamantane units increase rapidly for $n > 4$.

On fusing an additional adamantane unit onto an diamondoid consisting of n cages, two types of an $n + 1$ diamondoid can be formed. The first type is characterized by the chemical formula $C_{4n+6}H_{4n+12}$ and it can be obtained conceptually by replacing three axial hydrogen atoms bonded to the same cyclohexane ring. It can be assumed by addition of a trimethylenemethane group, resulting in the net addition of C_4H_4. Diamondoids of this type are called regular n-catamantanes, and for the same cage number n, they are isomeric (Balaban and Schleyer, 1978).

The other types of diamondoids, also named *irregular* catamantanes, result from the addition of C_xH_y with $x < 4$ and $y < 4$. An example of this is shown in Fig. 2.11b, which illustrates the formation of [1231]pentamantane from [123]tetramantane by the addition of C_3H_3. The necessary and sufficient conditions for a catamantane to be irregular is to have a code with at least two identical digits separated by any two other digits (not taking into account branches, that is, digits inside parentheses). For example, the sequences $a1bc1d\ldots$, $\ldots 2ab2c\ldots$, $ab3(b)cd3\ldots$, $\ldots abc4(a)b(d)a4d\ldots$, and $\ldots ab1(a)bc(1)d\ldots$ are examples of dualist codes of irregular catamantanes (here, the variables a–d stand for the digits 1 to 4).

For diamondoids up to and including tetramantane, the chemical formula of all possible isomers is $C_{4n+6}H_{4n+12}$. Starting with penta-

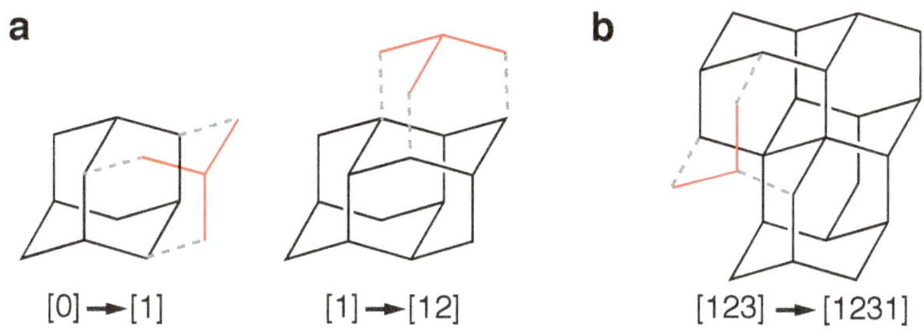

Figure 2.11 Formation of regular and irregular catamantanes. (a) Addition of a C_4H_4 group results in diamondoids with the chemical formula $C_{4n+6}H_{4n+12}$, for example, diamantane from adamantane and triamantane from diamantane. (b) For the case where the added fragment is C_xH_y with $x < 4$ and $y < 4$, (e.g., C_3H_3 for [123]tetramantane to [1231]pentamantane), so-called irregular catamantanes are formed.

mantane, $n = 5$, one can obtain an isomer of different molecular mass, [1231]pentamantane, with chemical formula $C_{4n+5}H_{4n+10}$. Continuing the series of diamondoids for $n \geq 6$, the number of isomers increases, with a new group having the generic chemical formula $C_{4n+2}H_{4n+6}$. With each increment of n by 1, new groups are added to the diamondoid family: $C_{4n+4}H_{4n+8}$ for $n \geq 7$, $C_{4n+1}H_{4n+4}$ for $n \geq 8$, $C_{4n-2}H_{4n}$ for $n \geq 9$, $C_{4n-5}H_{4n-4}$ for $n \geq 10$, and finally $C_{4n-3}H_{4n-2}$ for $n \geq 11$.

A list of these different diamondoid groups for $n = 1$ to 12 (dodecamantane) and their respective classes of molecular weights is summarized in Fig. 2.12.

2.4.4 Formula Partition Periodic Table

Extending the original work on the dualist graph-based classification of diamondoids, Balaban has proposed a formula partition periodic table of regular diamondoid valence isomers, that is, regular catamantanes that have the molecular formula $C_{4n+6}H_{4n+12}$ (Balaban, 2012).

As will be shown the following chapters, the position of the C atoms and their connectivity with other C atoms of the diamondoid structure influence the reactivity. While the dualist graph naming

Group \ n	1	2	3	4	5	6	7	8	9	10	11	12	...
1 $C_{4n+6}H_{4n+12}$	**136**	**188**	**240**	**292**	**344**	**396**	**448**	**500**	**552**	**604**	**656**	**708**	
2 $C_{4n+5}H_{4n+10}$					**330**	**382**	**434**	**486**	**538**	**590**	**642**	**694**	
3 $C_{4n+2}H_{4n+6}$						**342**	**394**	**446**	**498**	**550**	**602**	**654**	
4 $C_{4n+4}H_{4n+8}$							**420**	**472**	**524**	**576**	**628**	**680**	
5 $C_{4n+1}H_{4n+4}$								**432**	**484**	**534**	**588**	**640**	
6 $C_{4n-2}H_{4n}$									**444**	**496**	**548**	**600**	
7 $C_{4n-5}H_{4n-4}$										**456**	**508**	**560**	
8 $C_{4n-3}H_{4n-2}$											**534**	**586**	
⋮													

Regular, linear catamantanes

Catamantanes and perimantanes

Figure 2.12 Listing of molecular weights of diamondoids from adamantane ($n = 1$) up to dodecamantane ($n = 12$). The columns $n = 1$ to 12 denote the number of cages, while the rows 1 to 8 indicate different diamondoid groups that result from different possible cage arrangements. Group 1 diamondoids ($C_{4n+6}H_{4n+12}$) stand for regular, linear catamantanes, while diamondoids belonging to higher groups contain regular branched catamantanes, irregular linear and branched catamantanes, and irregular cyclic perimantanes (cf. Tables 2.1 and 2.2).

scheme proposed by Balaban and von Schleyer (Balaban and Schleyer, 1978) is more straightforward compared to the official IUPAC names, there are also some drawbacks.

For example, it is not easy to distinguish stereoisomers and the position of C and H atoms, and furthermore, it does not allow to deduce the number of secondary, tertiary and quaternary C atoms (i.e., whether a C atom is connected to two, three or four other C atoms) directly.

Therefore, in addition to the dualist graph based nomenclature of diamondoids, additional schemes have been developed to classify diamondoids.

Figure 2.13 displays the molecular structure of a series of diamondoids, with the colors of the C atoms indicating the number

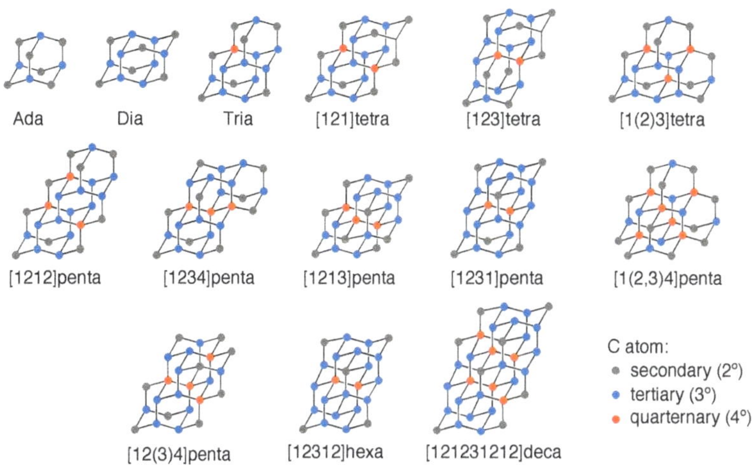

Figure 2.13 Molecular structures of diamondoids and C atoms coordination. The different types of C coordination are indicated by colors: linked to two C atoms, secondary (2°, gray); linked to three C atoms, tertiary (3°, blue); linked to four C atoms, quaternary (4°, red).

of other C atoms they are connected to: Four (quaternary, 4°), three (tertiary, 3°), or two (secondary, 2°). Both adamantane and diamantane only contain 3° or 2° C atoms. Then, starting with triamantane, one obtains diamondoids that also contain 4° C atoms, that is, it is the first diamondoid with a bulk-coordinated C atom. As will be shown in a subsequent chapter, the coordination of C atoms in the diamondoid framework has a large influence on their reactivity and is therefore paramount for applications involving their functionalization.

Based on the coordination number of C atoms, additional classification schemes of diamondoids have been developed. One consists in subdividing the chemical formula of the diamondoids into subgroups that indicate the C coordination number: $C_Q(CH)_T(CH_2)_S$. Here, the subscripts Q, T, and S indicate the number of 4°, 3°, and 2° C atoms, respectively. Table 2.3 lists the molecular formulae and the corresponding partitions for adamantane, diamantane, triamantane, and tetramantane.

Table 2.3 Partition table of regular catamantanes (chemical formula $C_{4n+6}H_{4n+12}$); the first row in the table is the chemical formula of the diamondoid separated into quaternary, tertiary, and secondary groups; the second row lists the number of the groups, and the third is the dualist degree partition (after (Balaban, 2012))

Number of cages n	Molecular formula	Zero branches C atoms Partition	One branch C atoms Partition
1	$C_{10}H_{16}$	$(CH)_4(CH_2)_6$ $0-4-6$ 0, 0, 0, 0	
2	$C_{14}H_{20}$	$(CH)_8(CH_2)_6$ $0-8-6$ 1, 1, 0, 0	
3	$C_{18}H_{24}$	$C(CH)_{10}(CH_2)_7$ $1-10-7$ 2, 1, 0, 0	
4	$C_{22}H_{28}$	$C_2(CH)_{12}(CH_2)_8$ $2-12-8$ 2, 2, 0, 0	$C_3(CH)_{10}(CH_2)_9$ $3-10-9$ 3, 0, 1, 0

2.5 Molecular Symmetry and Crystal Structures of Diamondoids

In this section, we give a brief overview over the molecular symmetry of diamondoids and the crystalline structures they adopt in the solid phase. As we will see in the chapter on vibrational spectroscopy of diamondoids, knowing the symmetry of the molecule can help in the interpretation of spectra and assigning peaks. Conversely, knowing the rules for Raman or IR active vibrational modes can help to guess the symmetry of the molecule under investigation. In addition, we also briefly list the known crystal structures that diamondoids accommodate in the solid phase.

2.5.1 Molecular Symmetry of Diamondoids

As will be explained in more detail, one convenient means to characterize diamondoids is by vibrational spectroscopy techniques, namely infrared (IR) and Raman spectroscopy. These techniques probe the different vibrational modes of the molecules. However, while both IR and Raman measurements are relatively easy to carry out, the interpretation of the spectra is not always straightforward. In order to understand in more detail the different types of spectral features that can be expected, to determine whether certain vibrational modes are either IR or Raman active, and to assign them to specific vibrational modes, it is helpful to know the symmetry groups and elements of the different diamondoids. Depending on the symmetry of a given diamondoid, the features in the spectra will change. In particular, diamondoids of higher symmetry will show less distinctive peaks compared to diamondoids having lower symmetry.

Here, we will not go too much into details of molecular symmetry and refer the interested reader to relevant textbooks of this topic. The main symmetry groups diamondoids belong to are C_1 (i.e., no symmetry element except identity, E), C_2, $C_{2\upsilon}$, C_{2h}, C_s, $C_{3\upsilon}$, D_{3d}, and T_d (Balaban and Schleyer, 1978; Dahl et al., 2003a).

As an example, we show the molecular structures and their symmetry elements of diamondoids ranging from adamantane to tetramantane in Fig. 2.14. The most symmetric diamondoids are adamantane (symmetry group T_d, cf. Fig. 2.14). The only another diamondoids belonging to the same symmetry group are [1(2, 3)4]pentamantane and [1231241(2)3]decamantane (the "adamantane-of-adamantanes" or "superadamantane"). All the other diamondoids are of lower symmetry, either completely lacking any symmetry element (C_1) or possessing only one axis of rotation C_2.

2.5.2 Crystal Structures of Diamondoids

In this part, we briefly discuss the crystalline structure of diamondoids in the solid phase. While from all the lower diamondoids, the crystal structures are known, in the case of higher diamondoids, only those that have been obtained in larger quantities could

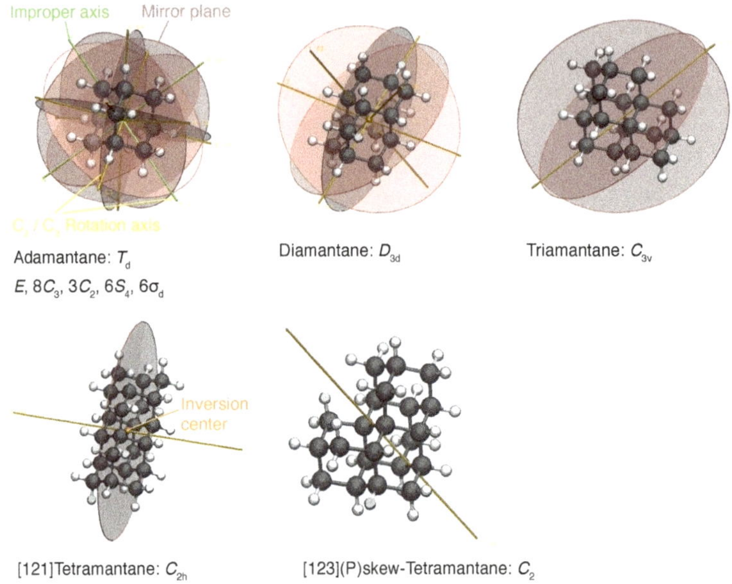

Figure 2.14 Examples of diamondoids, their molecular structures, and symmetry elements. Typically, diamondoids belong to point groups T_{d} (adamantane), $D_{3\mathrm{d}}$ (diamantane), C_{2v} (triamantane), $C_{2\mathrm{h}}$ ([121]tetramantane), and C_2 ([123](P)skew-tetramantane).

be characterized. Table 2.4 lists the known crystal structures of diamondoids and illustrates the crystal structures of adamantane and [1(2, 3)4]pentamantane in Fig. 2.15. Adamantane crystallizes in the face-centered-cubic (*fcc*) structure (space group $\mathbf{F\bar{4}3m}$), while [1(2, 3)]pentamantane is orthorhombic ($\mathbf{P2_1/m}$). In these crystalline structures, the diamondoid cages are only linked by van der Waals forces, but as will be shown in the next chapter, the regular molecular structure of especially the lower diamondoids leads to much higher thermals stability and melting points compared to alkanes of comparable molecular masses.

In the case of single-crystalline adamantane, two types of structures have been reported: $\mathbf{P\bar{4}2c}$ (triclinic) for temperatures below -65°C, and a disordered structure with space group **Fm3m**

Table 2.4 Listing of diamondoids with know crystal structure and corresponding space groups. The symbols in bold represent the space group, while the codes in parentheses are the Schönflies symbols

Name	Molecular formula	Dualist code	Crystal structure
Adamantane	$C_{10}H_{16}$	[0]	**Fm$\bar{3}$m** (O_h)
Diamantane	$C_{14}H_{20}$	[1]	**Pa$\bar{3}$** (T_h^6)
Triamantane	$C_{18}H_{24}$	[12]	**Fddd** (D_{2h}^{24})
Tetramantanes	$C_{22}H_{28}$	[121]	**P2$_1$/m** (C_{2h}^2)
		[123]	**P1** (C_1^1)
Pentamantanes	$C_{26}H_{32}$	[1212]	**P2$_1$2$_1$2$_1$** (D_2^4)
		[1(2, 3)4]	**Pnma** (D_{2h}^{16})
		[12(3)4]	**P2$_1$/m** (C_{2h}^2)
		[1213]	**P$\bar{1}$** (C_i^1)
Hexamantanes	$C_{26}H_{30}$	[12312]	**Pnma** (D_{2h}^{16})
Heptamantanes	$C_{30}H_{34}$	[121321]	**C2/m** (C_{2h}^3)

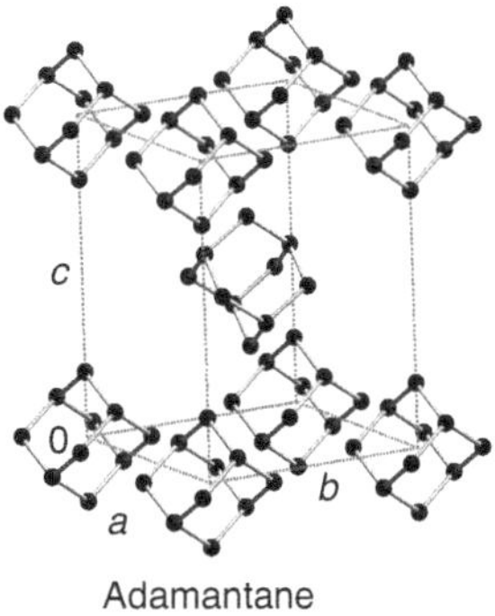

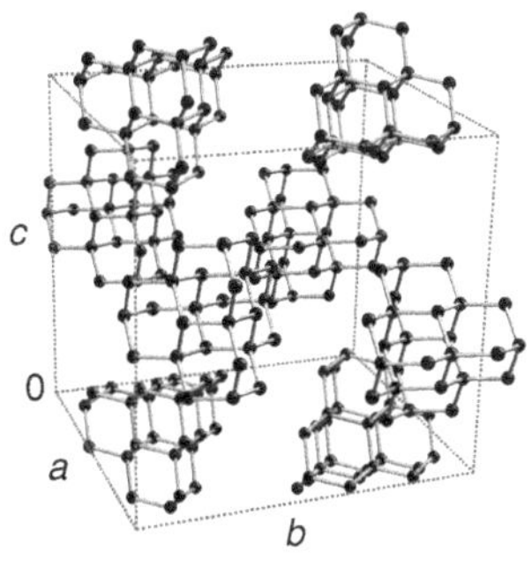

Figure 2.15 Crystal structures of adamantane (face-centered-cubic, space group **F$\bar{4}$3m**), and 1(2, 3)4pentamantane (orthorhombic, **Pnma**). The individual diamond cages in the crystals are only linked by relatively weak van der Waals forces.

(Nordman and Schmitkons, 1965), contrary to earlier reports that assigned the structure to be cubic (space group $\mathbf{F\bar{4}3m}$).

2.6 Differences between Diamondoids and Nanodiamonds

It might be in order to make a clear distinction between nanodiamonds and diamondoids. Although with increasing cage number, diamondoids become of similar size as nanodiamonds, (typical size 2–6 nm), it might be better to make a distinction between these two. Usually the term "nanodiamond" is used in different contexts and for a variety of structures that contain diamond nanocrystals, present, for instance, in interstellar dust and meteorites, diamond particles nucleated in the gas phase or on substrates, and nanocrystalline films (Shenderova et al., 2002). Typically, nanodiamonds may also possess a wide range of size scales, size distributions and also consist of mixtures containing other elements such as nitrogen and oxygen (i.e., doped nanodiamonds). In contrast, diamondoids "are chemically well defined, of high purity, and structurally well characterized" (Schwertfeger et al., 2008).

In particular, in the case of nanodiamonds, while their core is composed of sp^3-bonded carbon, their surface can be terminated by many different functional groups or consist of reconstructed surfaces formed by sp^2-hybridized graphitic networks (Raty and Galli, 2003). It has also been found that different fractions of detonation nanodiamond possessing different particle sizes also have varying surface compositions, namely different concentrations of hydrogen- or oxygen-containing groups (Larionova et al., 2006). Especially the fractions of smaller nanodiamonds with polar oxygen-containing groups (C—OH, C—O—C, C=O) lead to more facile dispersion in water.

In contrast, diamondoids have a well-defined structure and the sp^3 bonded carbon cages are terminated by H atoms or other functional groups. For this reason, diamondoids are also called diamond molecules or diamond hydrocarbons. Figure 2.16 illustrates the difference in the structure and surface terminations of diamond

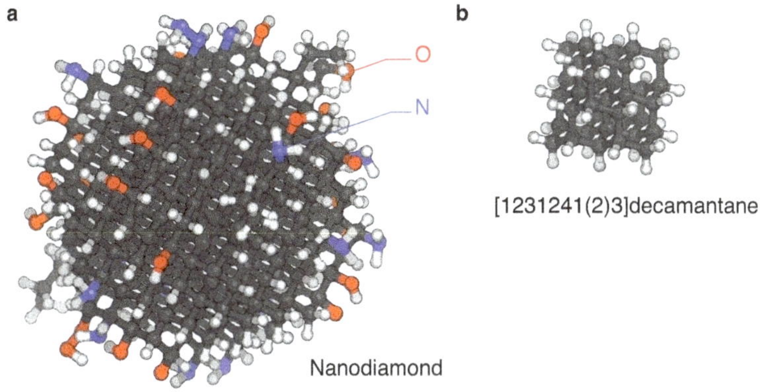

Figure 2.16 Comparison between structures nanodiamonds and diamondoids. (a) Example of nanodiamond consisting of a diamond core and surface that contains functional groups. (b) Structure of superadamantane; in contrast to nanodiamonds, the surface terminations are well defined and can consist of hydrogen atoms or other functional groups.

clusters and nanodiamonds on the one hand, and diamondoids on the other. Diamond clusters possessing not complete hydrogen terminations are also called bucky diamond (Raty et al., 2003) due to the resemblance of their reconstructed surface to the one of fullerenes.

Therefore, despite advances in their purification and functionalization, the major disadvantage of nanodiamonds remains their nonuniform size and size distribution (Schwertfeger et al., 2008). In contrast, as will be explained in more detail in the following sections, diamondoids are more similar to molecules, possessing well-defined structure, size, and composition.

As mentioned briefly in the previous section, most current applications of nanodiamond synthesis are based on detonation techniques, that was discovered in the 1960s (DeCarli and Jamieson, 1961; Dolmatov et al., 2004). However, once obtained, they have to be purified using strong (50%–99%) nitric acid at temperatures of 80°C–180°C. The chemical purification and oxidation steps are the most expensive and complex stages in the fabrication of nanodiamonds.

The growth of nanodiamonds by detonation synthesis is thought to occur by several mechanisms and in several stages. A recently proposed growth mechanism (Dolmatov et al., 2013) occurs in the following steps:

(1) Decomposition of the precursor and carbon source trinitrotoluene (TNT) into basic radicals and radical-like dimer C_2 and CO_3, decomposition of hexogen molecules into C_2 and of benzotrifuroxane molecules into C_2;
(2) Formation of cyclohexane from C_2 or immediately of radical adamantane molecules;
(3) Interaction of diamond-like core (adamantane radical) with methyl and other monocarbon radicals; and
(4) Growth of detonation nanodiamond particles like in a CVD process.

In the following chapter, we discuss the chemical and physical properties of diamondoids in more detail, before introducing current and possible future applications of diamondoids.

Chapter 3

Chemical and Physical Properties and Characterization of Diamondoids

This chapter builds on the previous one but goes more into the details of the molecular structure, chemical, thermodynamic, and physical properties of diamondoids. The first part of the chapter is devoted to the molecular structure of diamondoids, especially the exact lengths of the bonding angles and bond lengths between C and H atoms. We also discuss the electronic structure of the diamondoids more in detail, as this will permit us to understand the variation of the bandgaps of both H-terminated and H-functionalized diamondoids. The other parts are devoted to the thermodynamic properties, including solubilities both in organic solvents, gases, and supercritical fluids. Furthermore, we present the physical properties of diamondoids, that is, the variation of their bandgaps as a function of size and functionalization.

3.1 Chemical Properties

In this section, we give an overview of the chemical properties of diamondoids, including their molecular structure. In the first part

Diamondoids: Synthesis, Properties, and Applications
Sven Strauss and Kazuo Terashima

ISBN 978-981-4745-18-5 (Hardcover), 978-1-315-36466-7 (eBook)
www.panstanford.com

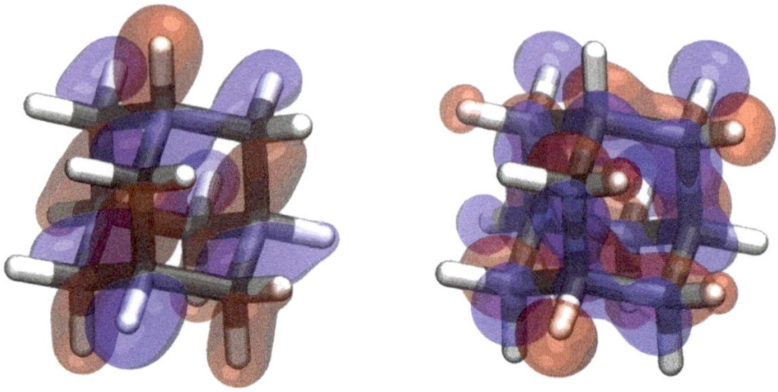

Figure 3.1 Molecular structure of adamantane and representation of its HOMO and LUMO. The red orbitals represent the negative part of the wavefunction and the blue the positive (the isosurface values have been set to ± 0.04).

of this paper, we will first discuss in more detail the electronic structure of diamondoids, in particular adamantane. In addition to the presentation of the chemical properties, we will describe the reactivities of diamondoids and the implication of these for the formation of new compounds.

Figure 3.1 presents a ball-and-stick model of adamantane, and the shape of the highest occupied molecular orbital (HOMO) and the lowest unoccupied molecular orbital (LUMO), as obtained by ab initio calculations. As can be seen in this figure, the HOMO of adamantane is mainly localized around the C—C and C—H bonds. Because of the hydrogen terminations of diamondiods, they are highly nonpolar. As will be shown in more detail in a later section, the solubility of diamondoids in typical organic solvents is limited.

So far, because with a few exceptions, especially larger diamondoids are not readily available, many of the chemical properties of diamondoids have not been measured directly, but rather have been predicted using computational chemistry approaches. Consequently, in the present text, we will also make frequent reference to such work from the scientific literature.

A detailed study on the structure and symmetry of diamondoids from adamantane to decamantane (adamantane-of-adamantane or superadamantane) (Shen et al., 1992) showed that for all levels of theories investigated (molecular mechanics, semi-empirical, and ab initio), that the C—C bonds in these molecules were longer than in bulk diamond (154.45 pm). Their simulations also showed very good agreement with experimental values of diamondoids that were available, that is, adamantane to triamantane.

Following the isolation and synthesis of the lower diamondoids in the 1950s and 1960s, their structures were investigated by X-ray diffraction (XRD) by several groups. The first study on the crystal structure of diamantane was carried out in 1965 (Karle and Karle, 1965).

In the case of triamantane, XRD analysis and Raman spectroscopy measurements indicated a change in the crystal structure, which was also confirmed by differential thermal analysis (DTA) (Jenkins and O'Brien, 1981). The onset of this transition was found to occur at about $20.5 \pm 0.5^{\circ}$.

Following its synthesis (cf. Chapter 6), the crystal structure of antitetramantane was investigated using XRD by (Roberts and Ferguson, 1977).

Recently, the effect of pressure on the crystal lattice of [121] Tetramantane has been investigated (Yang et al., 2014). The authors found that for pressures above 13 GPa, the monoclinic $P2_1/n$ structure changes to a triclinic P1 structure and the transition to the new structure was complete at a pressure of 16 GPa. The change in crystal structure displayed a large hysteresis and in addition, the appearance of a new phase with space group Pc, existing down to a pressure of 4 GPa, was observed when reducing the pressure. The phase change as a function of pressure was attributed to a change in the intermolecular packing, that is, the initially loosely packed unit cell became highly compacted at high pressure. The variation of the XRD spectra as a function of pressure up to 16 GPa is shown in Fig. 3.2.

While the structure of the lower diamondoids has been investigated both experimentally and numerically, the structure of larger diamondoids was long not possible because of insufficient computing power. In recent year, also the structural properties of larger diamondoids have been investigated, such as the one of cyclohexamantane (Richardson et al., 2006).

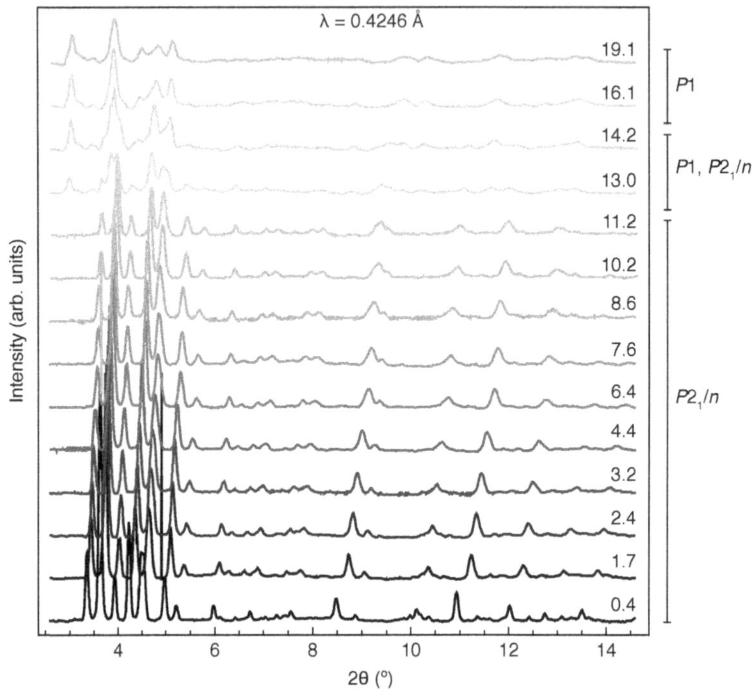

Figure 3.2 Variation of the crystal structure of [121]tetramantane as a function of pressure. The curves vary from $p = 0.4$ to 19.1 GPa. The lines on the right indicate the regions of the $P2_1/n$ and P1 phases. Between 13 and 16 GPa, P1 and $P2_1/n$ coexist, whereas for pressures above 16 GPa, only P1 remains. Data adapted from Yang et al. (2014).

3.1.1 Thermodynamic Properties

In this section, we discuss the main thermodynamic properties of diamondoids. Up to now, only a few diamondoids are available in sufficiently large quantities that allow such characterization, therefore the discussion focuses mainly on adamantane and diamantane.

In addition to these two diamondoid homologues, the standard enthalpies of formation of a few adamantoid compounds have also been investigated in detail. These included 1- and 2-methyladamantane; 2,2- and 1,3-dimethyladamantane; 1,3,5-trimethyladamantane; and 1,3,5,7-tetramethyladamantane (Steele and Watt, 1977). In general, the enthalpies of formation increase with increasing number of substitution from $-\Delta H_f^\circ = 242.7 \pm 1.7$ and 224.7 ± 1.7 kJ.mol^{-1} for 1- and 2-methyladamantane to 378.7 ± 3.8 kJ.mol^{-1} for 1,3,5,7-tetramethyladamantane.

The phase diagrams of adamantane and diamantane are displayed in Fig. 3.3. In the case of adamantane (Fig. 3.3a), three different states—solid, liquid, and gas—can be observed whereas for diamantane (Fig. 3.3b), three distinct different solid phases can be identified.

In addition to the different phases, both substances also exhibit a critical point.

At -65°C ($\approx$209 K), a phase transition occurs in adamantane (Nordman and Schmitkons, 1965). The phase change involves a transition from the face centered cubic room-temperature structure to a tetragonal structure, space group $P\bar{4}2_1c$, with $a = 660$ pm, $c = 881$ pm and $Z = 2$.

The low-temperature cell is related to the cubic cell by $a_{\text{tetra}} \simeq a_{\text{cubic}}/2$ and $c_{\text{tetra}} \simeq a_{\text{cubic}}$. A single-crystal X-ray study of the low-temperature phase at -110°C shows the arrangement of the molecules to be unchanged except for a 9° tilt about the c axis. The authors also found that at room temperature, the structure of adamantane is not well ordered and supports a structure that belongs to the space group $Fm3m$.

3.1.2 Chemical Stability

In the following, we will also briefly discuss the stability of diamondoids with respect to their size and in comparison to nanodiamondoids, and also as a function of their terminations, that is, the hydrogen content.

The structures of diamonds and the consequence of it on properties has been investigated since the earliest synthesis (Fort and Schleyer, 1964)

Since the discovery of diamond clusters of 3–5 nm in size from detonation synthesis and observation in meteorites, their stability with respect to bulk diamond has been investigated by several groups and suggested that such small diamond clusters could be more stable than graphite at ambient conditions (Badziag et al., 1990).

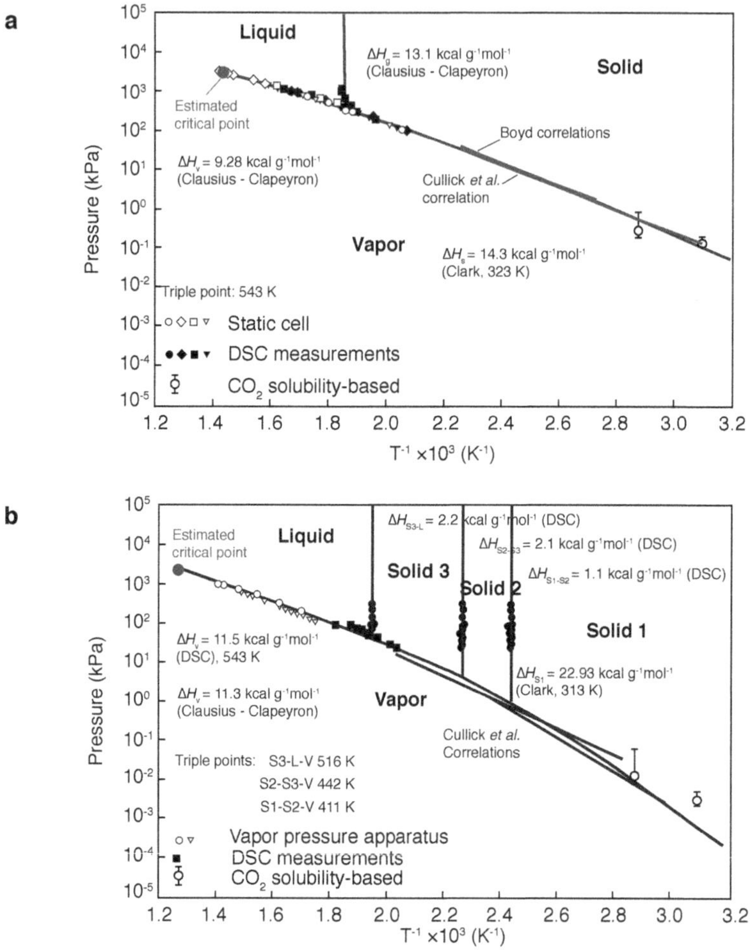

Figure 3.3 Phase diagrams of adamantane and diamantane, including the critical points. (a) Adamantane and (b) diamantane, which contains three different solid phases. Data adapted from Reiser et al. (1996).

The energy of formation E_{f} can be defined as:

$$E_{\mathrm{f}} = E_{\mathrm{tot}} - \mu_{\mathrm{C}} N_{\mathrm{C}} - \mu_{\mathrm{H}} N_{\mathrm{H}} - E_{\mathrm{vib}} \tag{3.1}$$

with E_{tot} the total energy, μ_{C}, μ_{H} and N_{C}, N_{H} are the numbers and chemical potentials of carbon and hydrogen atoms, and E_{vib} is the vibrational energy.

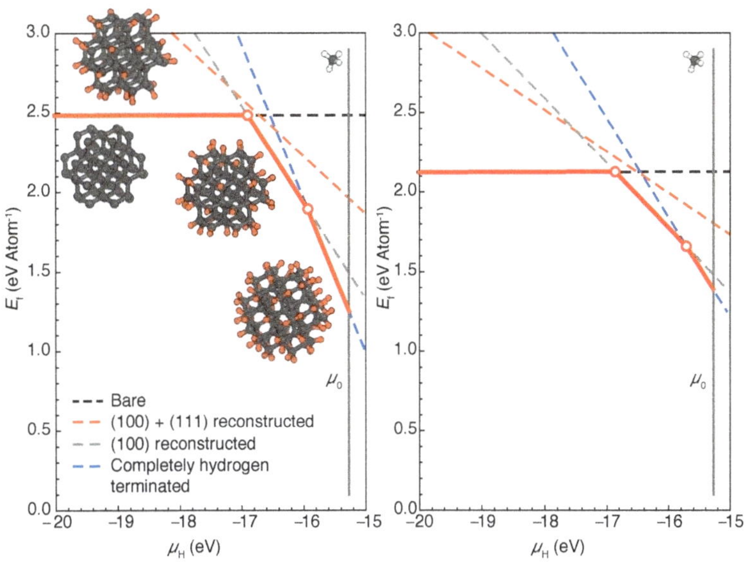

Figure 3.4 Comparison of the formation energy of diamond of different sizes and with different hydrogen coverage. (a) C_{66} diamond cluster. (b) Diamond cluster C_{275} (not shown). The energy is plotted as a function of the hydrogen chemical potential (μ_H), which can be attributed to different preparation conditions. Data adapted from Raty and Galli (2005). The line at $\mu_0 = -15.27$ eV is the formation energy of methane, whose molecular structure is also shown for size comparison. For better visibility, the hydrogen terminations of the nanodiamonds are colored in red. Note: The diamondoid structures depicted in the figure do not correspond to the actual, optimized structures.

The computed difference in formation energy per carbon atom (Δ) can be assumed to depend only on the number of surface carbon atoms. With this assumption, the difference in energy becomes

$$\Delta = \left(\frac{E_{\text{f, bare}}}{N_C}\right)^{3/2} - \left(\frac{E_{\text{f, stable}}}{N_C}\right)^{3/2} \tag{3.2}$$

It has be found (Raty and Galli, 2005), that Δ is almost linear with the particle diameter, as shown in Fig. 3.4, where the variation of E_f as a function of the chemical potential of hydrogen, μ_H is presented.

Figure 3.5 shows the difference in the formation energy between nanodiamonds possessing fully reconstructed surfaces Δ (in eV$^{1.5}$/Carbon), and the most stable structures with hydrogen

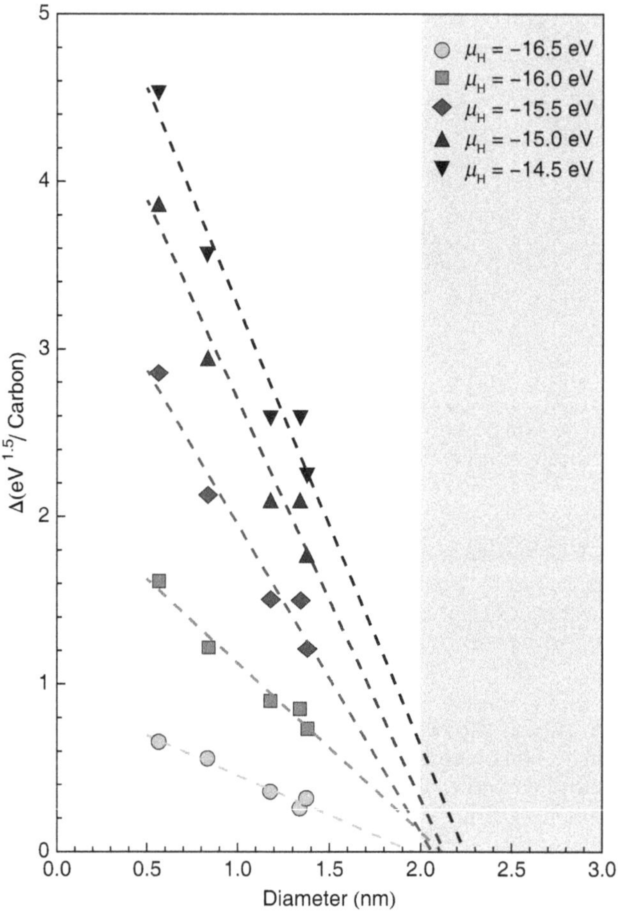

Figure 3.5 Variation in formation energies of nanodiamonds as a function of surface reconstruction and size. The differences in formation energies to the power 3/2, Δ, are between those of nanodiamonds possessing fully reconstructed surfaces and the most stable nanodiamond structure, for a given value of the hydrogen chemical potential μ_H. The shaded gray area for particle sizes between 2 and 3 nm indicates the domain where nanodiamonds with bare unreconstructed surfaces become more stable than those with hydrogen at their surface. Data adapted from Raty and Galli (2005).

terminations. As can be seen from the figure, the difference decreases with diamond size and reaches zero for sizes in the range between 2 and 3 nm.

These calculations show that for sizes above 3 nm, nanodiamonds with reconstructed surfaces are more stable than nanodiamonds possessing complete hydrogen terminations.

The size of nanodiamonds and their stability is also crucial for biological applications. By performing MD simulations of annealing in vacuum and in the presence of oxygen, it was found that diamond clusters of cubic or cuboctahedral shapes with over ~250 atoms can survive long-term annealing up to temperatures of 1500 K. The primary role of oxygen is the saturation of surface carbon atoms and the etching of the graphitic part, that is, to increase the sp^3 ratio. The results showed that {100}-faceted nanodiamonds with sizes as small as 1 nm can be kinetically stable up these temperatures (Kaviani et al., 2013).

Figure 3.6 shows computationally obtained relative stabilities (ΔH_{298}) of tertiary diamondoidyl cations, which are obtained from the reaction of the 1-adamantyl cation ($1 - Ad^+$) and a given

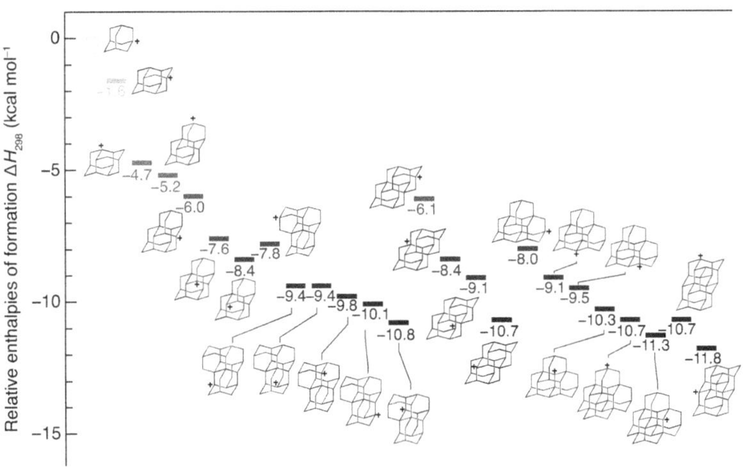

Figure 3.6 Relative stabilities of tertiary diamondoidyl cations in comparison to the 1-adamantyl cation. The values of ΔH_{298} are obtained from Eq. 3.3. Data adapted from Fokin et al. (2005).

diamondoid (Fokin et al., 2005):

$$1 - \mathrm{Ad}^{+} + \mathrm{diamondoid} \longrightarrow \mathrm{AdH} + \textit{tert}\text{-diamondoidyl}+ \qquad (3.3)$$

3.1.3 Solubility of Diamondoids in Gases, Organic Solvents, and Supercritical Fluids

The pressure-temperature conditions in the natural gas production or oil exploration often lead to the formation of solids, which then clog the pipes, which leads to delays and ultimately production losses. Therefore, for the exploration of diamondoids from natural gas or oil reservoirs or for their further treatment—separation and purification—by organic chemistry, knowing how well they are soluble in a given fluid is essential. This section contains an overview of most of the currently available data on the solubility of diamondoids in gases, organic solvents, and supercritical fluids.

3.1.3.1 Solubility in organic solvents

Diamondoids often clog pipes in oil or gas reservoirs and therefore, to mitigate this problem, it is essential to know the solubility limits of diamondoids in various fluids and for different temperature and pressure conditions.

Several research groups have studied the solubility of diamondoids in various organic solvents, and both atmospheric and high-pressure conditions, including the supercritical state. For example, Smith and Teja (Smith and Teja, 1996) have investigated the solubility of adamantane in several organic solvents. Figure 3.7 displays the variation of the solubility y_2 of adamantane and diamantane as a function of solvent density for CO_2, methane (CH_4), and ethane (C_2H_6).

The solubility of several diamondoids and derivatives of diamantane in typical organic solvents has also been studied (Chan et al., 2008). The mass fraction solubilities (S) of diamantane, triamantane, and tetramantane as a function of temperature are listed in Table 3.1 (in the case of tetramantane, the solubility values are those of four isomers).

In general, saturated diamondoids are more easily dissolved in nonpolar solvents. For diamondoids that contain polar groups,

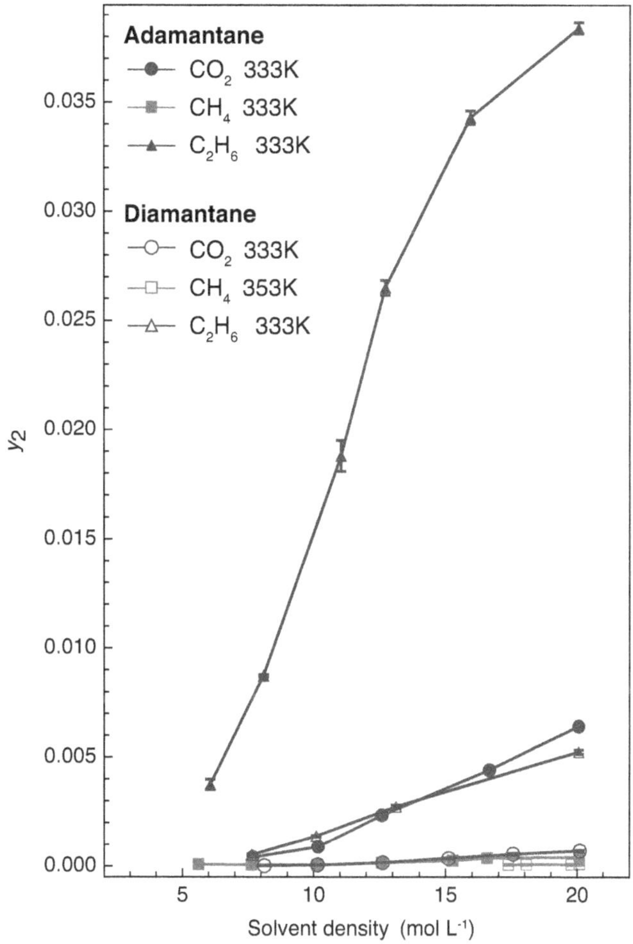

Figure 3.7 Solubilities y_2 (solid mole fraction) of adamantane and diamantane in CO_2, CH_4, and C_2H_6 as a function of solvent density. Data compiled from Smith and Teja (1996) and Swaid et al. (1985).

for example, 1-hydroxyl diamantane or 1,6-dihydroxyl-diamantane, they can also be dissolved in polar solvents.

In case the melting temperature T_m, the enthalpy of fusion ΔH_{fus}, and the activity coefficient γ_2 of the solute are known, then the

Table 3.1 Experimental mass fraction solubility, S, of diamantane, triamantane, and tetramantane in cyclohexane, ethyl acetate, toluene, acetone, heptane, and 1-pentanol as a function of temperature T. Data adapted from Chan et al. (2008)

	100 w					
T (K)	Cyclohexane	Ethyl acetate	Toluene	Acetone	Heptane	1-Pentanol
			Diamantane			
283.15	4.336	0.555	2.790	0.274		
293.15	5.430	0.820	3.976	0.412		
303.15	6.827	1.191	5.132	0.633		
313.15	9.144	1.495	7.148	0.905		
323.15	12.259	2.413	9.520	1.425		
333.15	14.945					
343.15	19.022					
			Triamantane			
273.15	5.396	0.396	3.802		2.842	0.600
298.15	9.553	1.010	7.642		6.503	1.272
323.15	15.492	2.120	14.314		10.651	2.725
348.15	26.012	5.175	26.152		17.564	5.436
			Tetramantane			
273.15		0.216	3.032		2.360	0.474
298.15	2.968	0.412	5.885		3.733	0.958
323.15	5.517	1.327	9.748		6.942	2.006
348.15	8.992	3.980	14.511		14.219	4.118

solubility can be estimated from

$$\ln(\gamma_2 x_2) = -\frac{\Delta H_{\text{fus}}}{R}\left(\frac{1}{T} - \frac{1}{T_{\text{m}}}\right) \tag{3.4}$$

where x_2 is the mol fraction of the solute and $R = 8.314462$ J.mol^{-1}.K^{-1} is the universal gas constant.

The activity coefficients γ_2 of the solute can be estimated from a model such as the nonrandom two-liquid (NRTL) model:

$$\ln(\gamma_2) = x_1^2\left[\tau_{12}\left(\frac{G_{12}}{x_2 + x_1 G_{12}}\right)^2 + \frac{\tau_{12} G_{21}}{(x_1 + x_2 G_{21})^2}\right] \tag{3.5}$$

where $G_{12} = \exp(-\alpha_{12}\tau_{12})$, $G_{21} = \exp(-\alpha_{12}\tau_{21})$, and τ_{12} and τ_{21} are interaction parameters.

Solubilities of adamantane and diamantane have also been measured in pressurized hot water up to 5 MPa, at temperatures between 313 K and the solid-liquid-vapor triple point of the solute (Karasek et al., 2008). Figure 3.9 illustrates the solubilities of adamantane, diamantane, and a series of polycyclic aromatic hydrocarbons in aqueous solutions as a function of temperature.

There have also been a few reports on the solubilities of diamondoids in gaseous systems. However, because of the very small solubilities (on the order of $\sim 10^{-4}$), measuring the concentrations is very difficult and there are only a limited amount of experimental data available. For adamantane and diamantane, a thermodynamic data assessment test has been performed (Eslamimanesh et al., 2011) to determine their solubilities in supercritical CO_2 and ethane.

3.1.3.2 Solubility in gases and supercritical fluids

Because of their high thermal stability and low volatility, diamondoids can be concentrated in the exploration of gas or oil reservoirs and lead to clogging of pipes. To understand the conditions under which diamondoids solidify and start clogging piping systems, the solubility of diamondoids in high-pressure and supercritical fluids, especially CO_2, has been investigated (Swaid et al., 1985).

Figure 3.8 displays the variation of the solubility of adamantane in supercritical CO_2 ($scCO_2$), as a function of temperature, pressure, and density.

As has been explained above, diamondoids cannot be readily dissolved in polar solvents, unless the diamondoids contain polar functional groups. Because of the low solubility, data in polar solvents is difficult to measure, but the aqueous solubilities of the two smallest diamondoids adamantane and diamantane have been investigated by Karásek et al. (2008).

Figure 3.9 shows the solubilities of adamantane, diamantane, and three polycyclic aromatic hydrocarbons (PAHs) in aqueous solutions. The PAHs chosen for comparison are naphthalene ($C_{10}H_8$), who has the same number of carbon atoms as adamantane, and anthracene and phenanthrene, who both have the same number of carbon atoms as diamantane. As expected, the solubility of diamantane is much

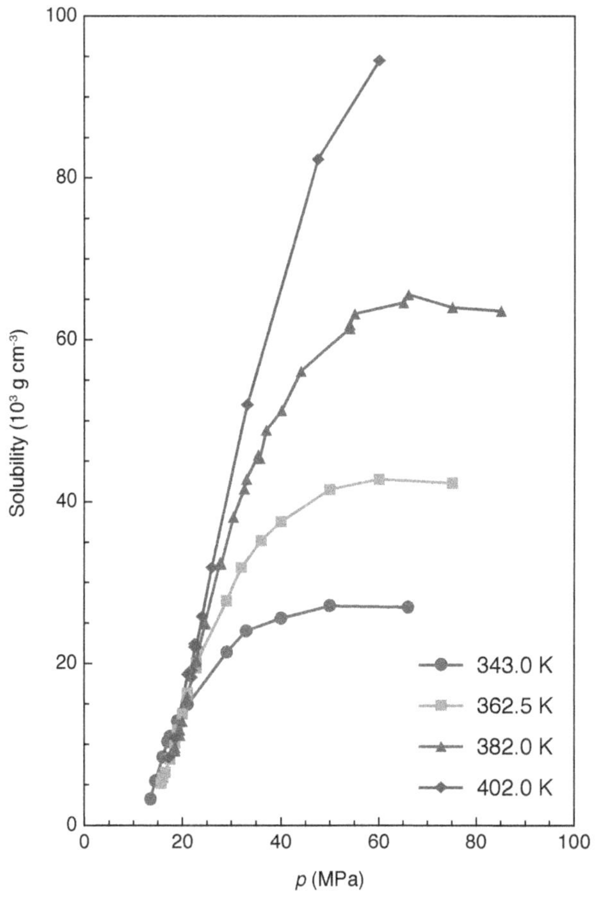

Figure 3.8 Variation of the solubility of adamantane in supercritical CO_2. The data shows the variation in the pressure range from ~10 to ~90 MPa and between temperatures of 343, 362, 382, and 402 K (these values correspond to $p/p_{crit} = 1.36–12.20$ and $T/T_{crit} = 1.13–1.32$). Data adapted from Swaid et al. (1985) and Kraska et al. (2002).

lower compared to adamantane. The solubilities can be fitted by the following equation:

$$\ln x_2 = a_1 + a_2(T_0/T) + a_3 \ln(T/T_0) \tag{3.6}$$

where $T_0 = 298.15$ K, and a_i, $i = 1, 2, 3$ are fitting coefficients.

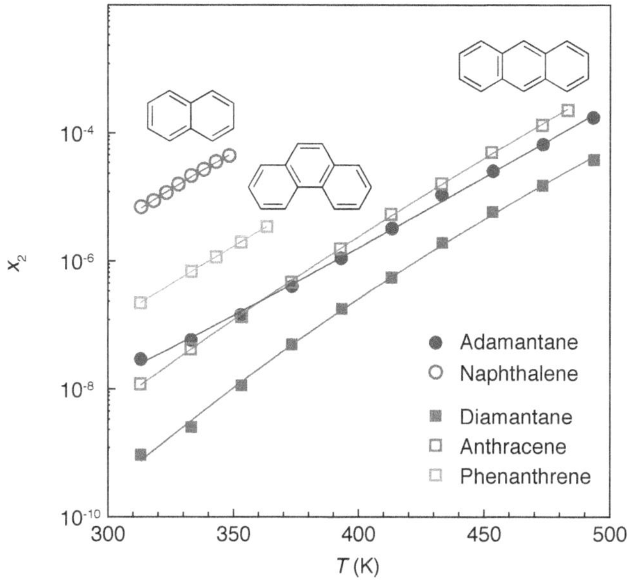

Figure 3.9 Molar solubilities of diamondoids and polycyclic aromatic hydrocarbons in aqueous solutions. The graph shows the variation of the molar solubility x_2 of adamantane, diamantane, and the polycyclic aromatic hydrocarbons naphtalene ($C_{10}H_8$), anthracene ($C_{14}H_{10}$), and phenanthrene ($C_{14}H_{10}$), whose molecular structures are also indicated. The lines are the best fits of the solubility according to Eq. 3.6. Data adapted from Karasek et al. (2008).

Between temperatures of $T = 254$ and 543 K, the sublimation pressure of the solid phase Ada(I) is (Bazyleva et al., 2011):

$$\ln(p/\text{kPa}) = 50.9129 - 8494.5(T/\text{K})^{-1} - 4.6395\ln(T/\text{K}) \quad (3.7)$$

3.1.4 Biocompatibility and Toxicity

Because of their stable structure and inertness, in recent years, nanodiamonds have attracted increasing attention for potential applications in biotechnology and medicine.

At the time of writing of this book, to the best of our knowledge, there were no in vitro or in vivo studies on the toxicity of diamondoids. However, since recently, nanodiamonds have become an attractive alternative to other nanoparticles (e.g., functionalized metallic nanoparticles) for labeling and drug delivery, studies investigating possible toxicity effects of nanodiamonds in organisms have increased (Krueger, 2008, 2011; Krueger and Lang, 2012).

Therefore, because from a structural point of view, diamondoids are nanodiamonds with sizes below 2 nm, one could argue that at least for diamondoids that are of comparable size to nanodiamonds, similarities in the behavior of organs and organisms exist.

For biotechnology applications, nanodiamondoids have become increasingly popular as fluorescent probes and for drug carriers, because of their exceptional stability and inertness. Moreover, there is no known toxicity of diamondoids. The reason why in the current understanding, diamondoids are nontoxic is that the metabolism of people allows diamondoids to be removed.

As will be shown in Chapter 4, nanodiamonds and diamondoids show promise for applications in medicine and biotechnology (Schrand et al., 2009; Wang et al., 2003). However, this brings up the question whether nanodiamonds, especially when used for the treatment of humans, are toxic or not (Schrand et al., 2007b). In a study that investigated the toxicity of several carbon nanomaterials, nanodiamond, carbon black, multiwall carbon nanotubes and single-wall carbon nanotubes found that the biocompatibility was highest for nanodiamonds (Schrand et al., 2007a).

Over the last few decades, several studies of diamondoids and their effect on organisms have been conducted. So far, these studies have been mainly realized on cells or animals. For example, the metabolism of several diamondoids and diamondoid derivatives in rat livers has been investigated (Hodek et al., 1988).

Applications of nanodiamonds for bioapplications: In the following, we discuss possible applications of nanodiamonds for bioapplications. So far, with the exception of diamondoid derivatives that have been used in pharmaceutics, there have been no investigations on the effect of diamondoids on cells, living tissue, etc. Moreover, as will be shown later, at present, there diamondoids are not available in large quantities, and therefore, the discussion has to be limited to nanodiamonds. But we hope, that in the future, that current applications of nanodiamonds can be extended to diamondoids.

The accumulation of nanodiamonds (NDs) with sizes of 50 nm has been investigated (Yuan et al., 2010). Using mouse samples, they found that the NDs accumulated mainly in the liver, lung, and spleen, and that the level of nanodiamonds amassed in these organs were kept constant even after 28 days.

3.2 Physical Properties

In this section, we will discuss the main physical properties of diamondoids. Specifically, we will discuss the electronic structure and how it changes when the size of the diamond increases. We also will present some approaches for tuning the bandgap in diamondoids, by the inclusion of atoms inside their cages, or by functionalization. Different spectroscopic methods can be used to characterize diamondoids and we present the main characteristics of vibrational spectra acquired on diamondoids and their main features. Finally, we also briefly discuss the main characteristics of mass spectra acquired on diamondoids.

Diamondoids can assemble to larger, macroscale crystals, bonded by weak van der Waals forces, forming structures that resemble macroscale diamonds. However, it is at the intramolecular sp^3 carbon framework level, not the intermolecular level, that diamond-like properties, for example, strength, rigidity, and stability, emerge (Dahl et al., 2003a).

3.2.1 Electronic Properties

In this section, we take a closer look at how the diamondoid structure—size, the substitution of carbon atoms in the carbon framework, by attaching functional groups, the positions where functional groups are attached—influences their electronic properties. We also draw comparisons between the electronic structures of diamondoids, nanodiamonds, and bulk diamond.

To assess the band structure of diamondoids, a lot of work has been done using density functional theory (DFT), and we will compare these works to available experimental data whenever possible.

3.2.1.1 The band structure of diamondoids

In order to understand the band structure of diamondoids, we will briefly compare them to the band structure of bulk diamond. Figure 3.10a shows the band structure of crystalline adamantane, diamantane, and triamantane.

The experimental bandgap of diamond is $\Delta E_{\mathrm{gap}} \approx 5.46$–$5.5$ eV and it is an indirect semiconductor.

To measure the bandgaps in diamondoids, essentially two techniques have been employed up to now: X-ray absorption

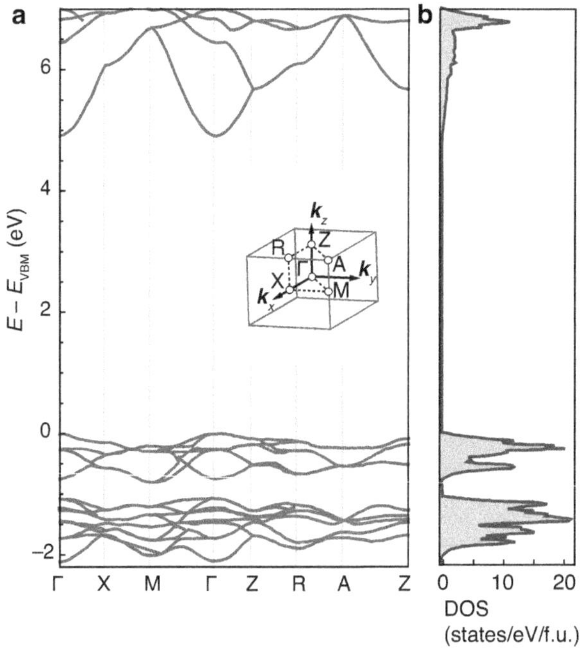

Figure 3.10 Electronic band structures of crystalline adamantane. (a) Band structure of crystalline adamantane along the high symmetrical momentum points. The inset shows the Brillouin zone for the tetragonal primitive cell with symmetry points labeled according to the standard notation. (b) The density of electronic states (DOS) in crystalline adamantane. The energy is plotted with respect to the valence band maximum (E_{VMB}). Data adapted from Sasagawa and Shen (2008).

spectroscopy (XAS) and soft X-ray emission (SXE) spectroscopy. XAS of carbon 1*s* levels can be used to assess the unoccupied states and can be used to estimate the relative energy position of the lowest unoccupied orbitals. In contrast, SXE allows measuring the highest occupied sp^3-hybridized states in diamondoids. In SXE, the lower energies and their measurements suggest an increase in the bandgap for decreasing diamondoid size. To estimate the HOMO-LUMO gap, DFT simulations have been conducted at different levels of theory. However, in some cases, the differences between the various adopted methods were quite large.

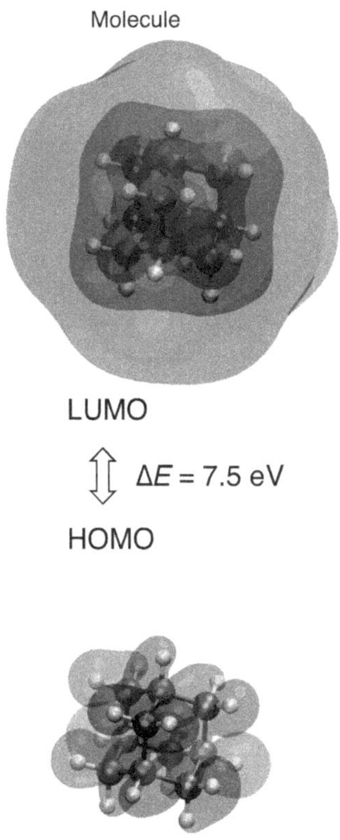

Figure 3.11 Orbitals of molecular and crystalline adamantane: HOMO (bottom) and LUMO (top) of adamantane.

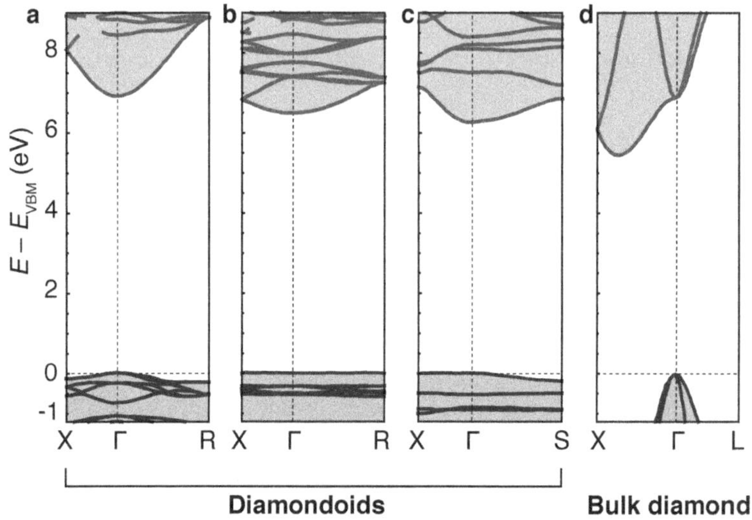

Figure 3.12 Comparison of bandgaps of crystalline diamondoids and bulk diamond. (a) Adamantane, (b) diamantane, (c) triamantane, and (d) bulk diamond. The energy dispersion relationship is shown around the origin of the momentum space (Γ point) for the crystalline structures of the respective diamondoids. Data adapted from Sasagawa and Shen (2008).

3.2.1.2 Effect of diamondoid size on bandgap

It has been shown that the bandgap of diamondoids can be tuned by changing their size or by the incorporation of different particles in the interior.

Figure 3.11 shows again an example of the HOMO and LUMO of adamantane. Compared to the isolated molecule, in the case of solid adamantane, the HOMOs are mainly situated around the C–C and C–H bonds.

Figure 3.12 displays the change of the band structure between the lower diamondoids adamantane, diamantane, and triamantane in comparison with the one of bulk diamond. The lower diamondoids possess a direct bandgap (indicated by the line at the Γ point), whereas in bulk diamond, the bandgap is indirect.

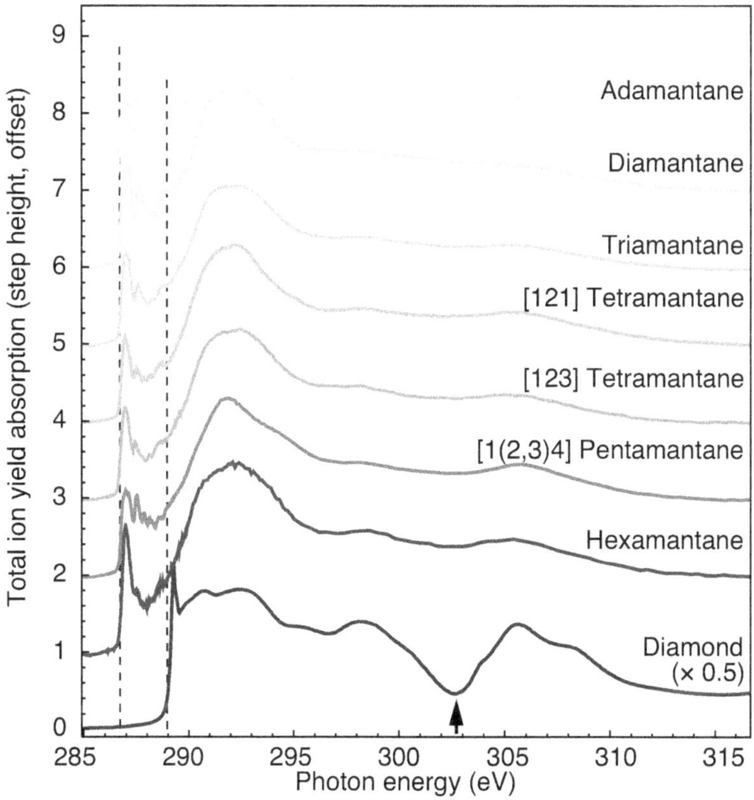

Figure 3.13 X-ray absorption spectroscopy of diamondoids in comparison to diamond. For the sake of a better comparison, the intensities of the curve of diamond have been multiplied by a factor of 0.5. Data adapted from Willey et al. (2005).

The X-ray absorption spectra of various diamondoids in comparison to diamond is illustrated in Fig. 3.13.

Figure 3.14 shows the variation of the normalized photo-ion yield for diamondoids ranging from adamantane to pentamantane.

Therefore, the bandgap of diamondoids can be modified by several approaches:

(1) Modification of the size
(2) Functionalization
(3) Inclusion of heteroatoms inside the diamondoid cage

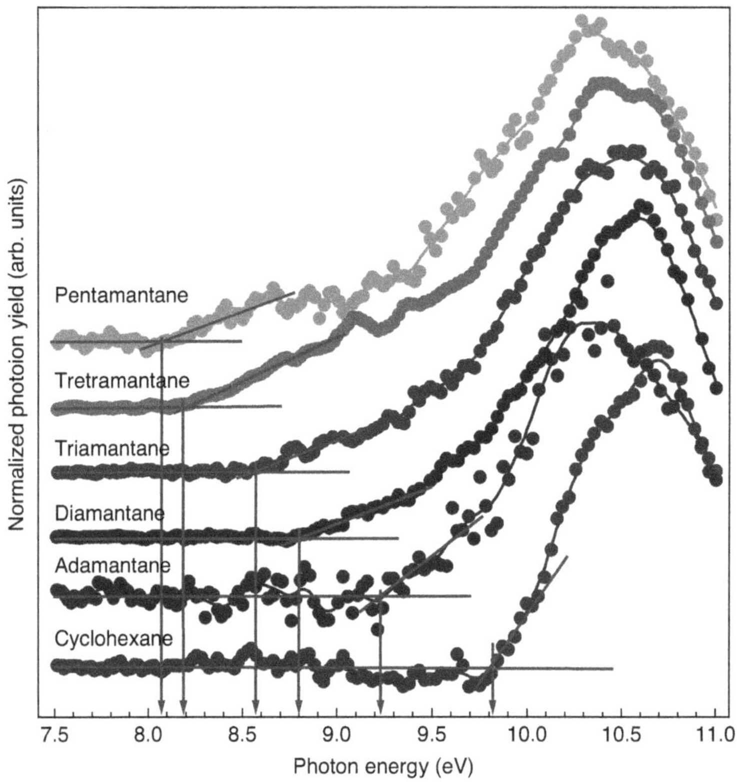

Figure 3.14 Photo-ion yield for different diamondoids and cyclohexane. The photo-ionization level is found by the intersection of the horizontal line and the linear fit to the increase of the photo-ion yield. Ada: adamantane; Dia: diamantane; Tria: triamantane; Tetra: [121]tetramantane; Penta: [1234]pentamantane. Data adapted from Lenzke et al. (2007).

The variation of the bandgap of diamondoids as obtained simulations is illustrated in Fig. 3.15, which shows the variation of the bandgap for both icosahedral and octahedral diamondoids. By increasing the size from adamantane ($C_{10}H_{16}$, $n = 1$) to $C_{286}H_{144}$ the bandgap decreases from about 9.5 eV to about 6.8 eV.

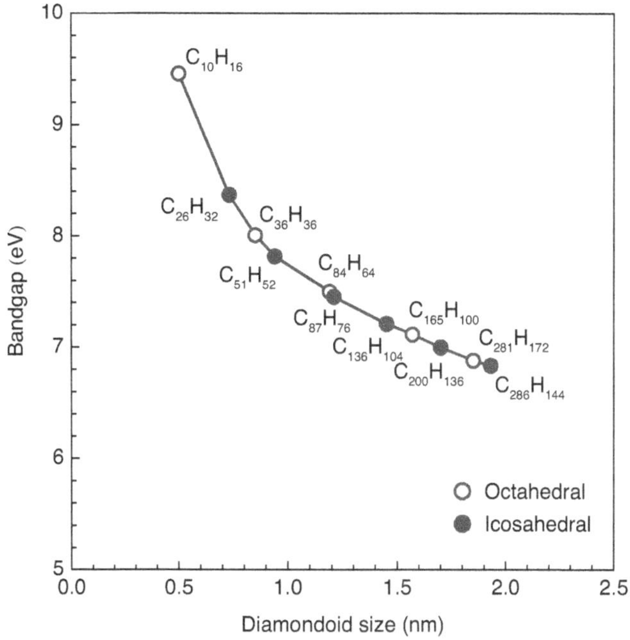

Figure 3.15 Variation of the bandgap of diamondoids as a function of diamondoid size. The bandgaps, estimated from density functional theory simulations, are shown for $C_{10}H_{16}$ to $C_{286}H_{144}$, for both diamondoid clusters with icosahedral and octahedral symmetry. Data adapted from Fokin and Schreiner (2009).

3.2.1.3 HOMO-LUMO variation by functionalization of diamondoids

In addition to the size of the diamondoids, the bandgap of diamondoids can be modified by partially replacing the H terminations by other atoms or molecules.

Figure 3.16 to 3.18 show valence spectra of diamantane and functionalized diamantane acquired by photoelectron spectroscopy, which illustrate the effect of different functionalization strategies.

Figure 3.19 displays the variation of the orbital energies of sulfur-doped diamondoids (thiamantanes).

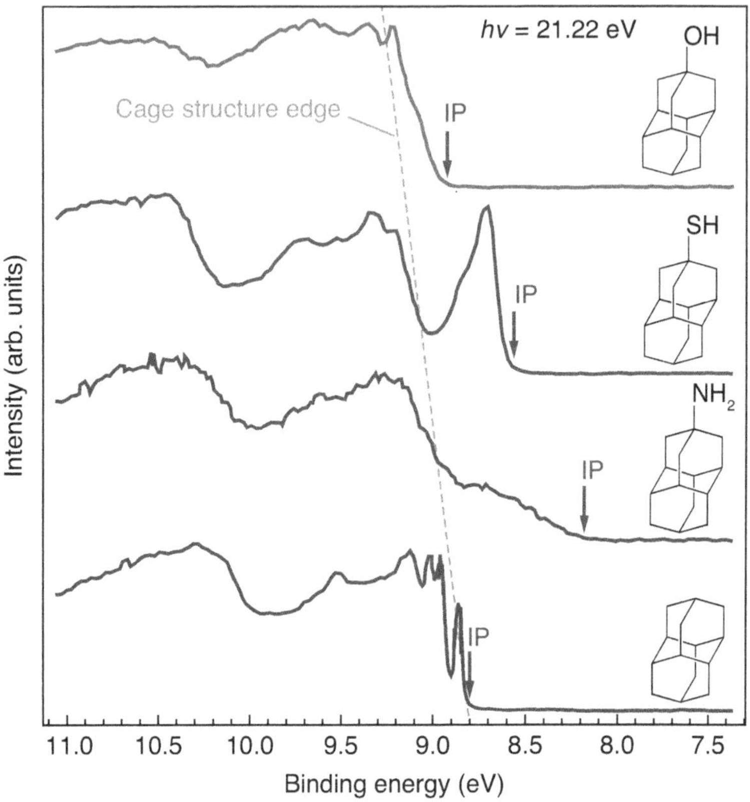

Figure 3.16 Valence spectra of diamantane and functionalized diamantane. The arrows indicated the ionization potentials (IPs), and the dashed line indicates the approximate position of the cage structure edge. Data adapted from Rander et al. (2013).

3.2.1.4 Variation of the HOMO-LUMO gap by the inclusion of small atoms inside diamondoids

In addition to bandgap tuning by functionalization, that is, the replacement of the hydrogen termination by functional groups, another possibility that has been investigated by Marsusi et al. (Marsusi and Mirabbaszadeh, 2009) is by including different small atoms (Li, Be, He, and Ne) at the interior of the diamond cage. Figure 3.20 shows the molecular structures of the first three diamondoids and an included atom.

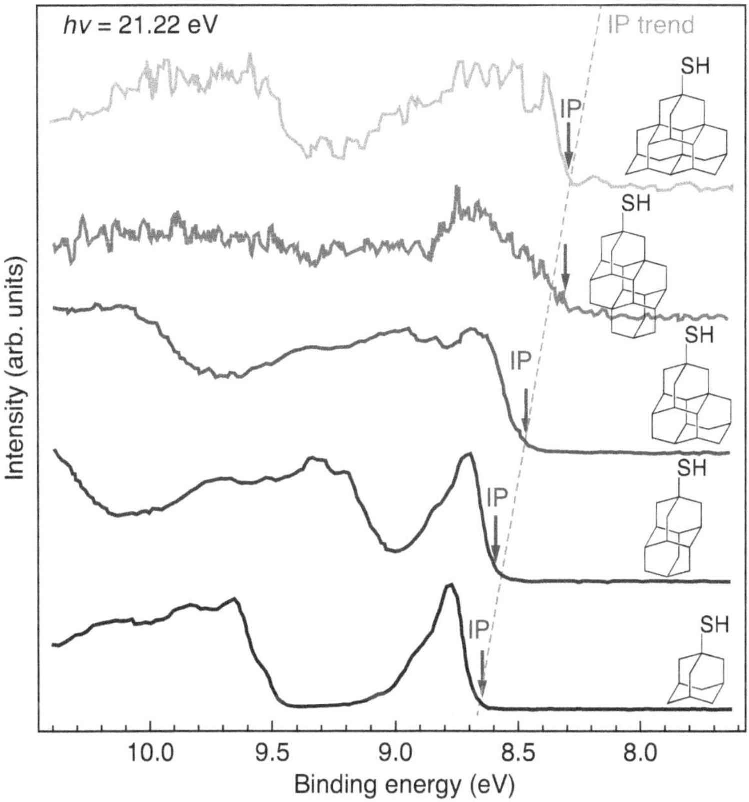

Figure 3.17 Valence spectra of diamondoid thiols. The arrows indicated the ionization potentials (IPs), and the dashed line indicates its trend. Data adapted from Rander et al. (2013).

Figure 3.21 shows the variation of the HOMO-LUMO gaps calculated for different structures similar to those shown in Fig. 3.20. Specifically for adamantane, the effect of encapsulating transition metal atoms inside its cage has been investigated in detail (Garcia et al., 2011).

In addition the influence of a single thiol group on the electronic and optical properties of the smallest diamondoid adamantane has also been investigated (Landt et al., 2010).

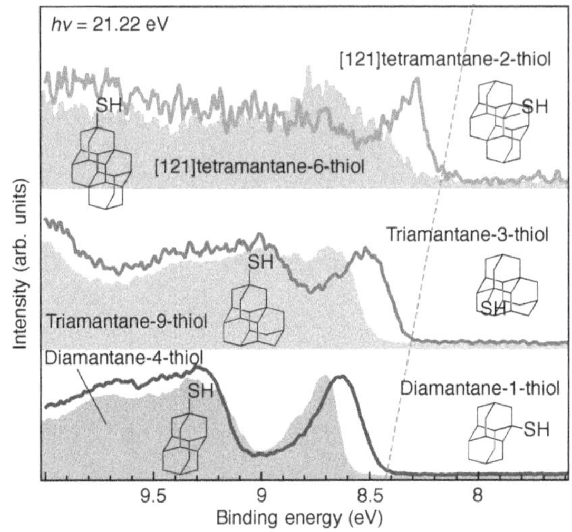

Figure 3.18 Variation of valence spectra of diamondoid thiols as a function of position. The curves show the spectra for the thiol derivates of diamantane, triamantane, and [121]tetramantane, for both apical and median positions. Data adapted from Rander et al. (2013).

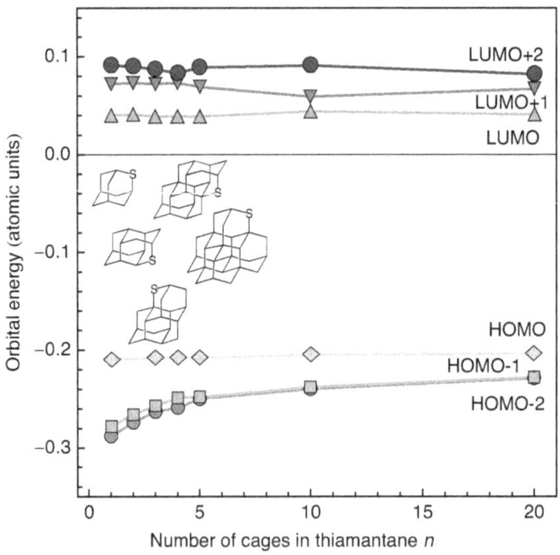

Figure 3.19 Variation in the orbital energies of the unoccupied and occupied states with increasing size of S-doped diamondoids (thiamantanes). The calculations were conducted at the B3PW91/6-31G(d, p) level of theory. Data adapted from Fokin and Schreiner (2009).

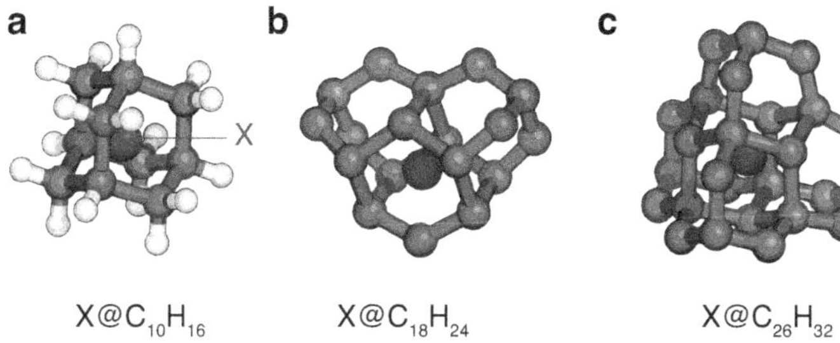

Figure 3.20 Schematic of molecular structures with atoms included inside diamondoid cages. (a) Adamantane, (b) triamantane, and (c) [1(2,3)4]pentamantane. The atom inserted inside the cage (noted X) investigated were Li, Be, He, and Ne. For clarity, the atoms are not drawn to scale and for X@triamantane and X@[1(2,3)4]pentamantane, the hydrogen atoms are not shown.

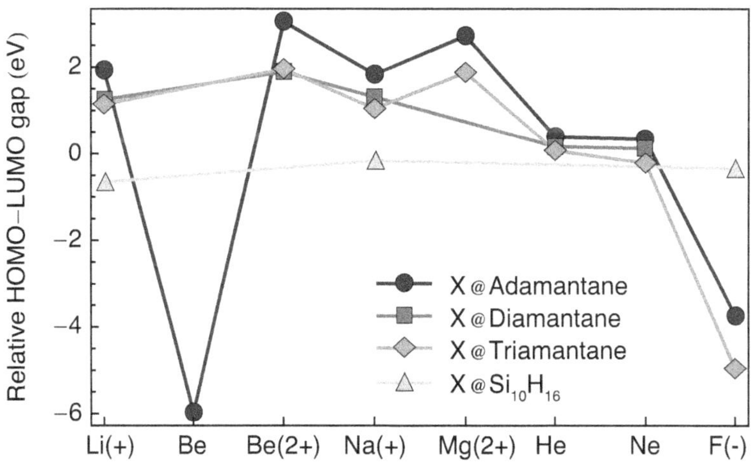

Figure 3.21 Relative HOMO-LUMO gaps of doped diamondoids consisting of different inclusion atoms calculated at the B3LYP/6-31G(d,p) level of theory. These structures represent possible spin singlet complexes $X@C_{10}H_{16}$, $X@C_{18}H_{24}$, and $X@C_{26}H_{32}$. The data is shown relative to the corresponding single-cage gaps, along with the relative HOMO-LUMO gap of $Si_{10}H_{16}$ taken from Pichierri (2006). Data adapted from Marsusi and Mirabbaszadeh (2009).

3.3 Optical Properties

3.3.1 Vibrational Spectroscopy of Diamondoids

3.3.1.1 Infrared spectroscopy

The vibrational spectra of adamantanes have been investigated in detail by Ramachandran et al. (Ramachandran and Manogaran, 2006, 2007), while the infrared spectrum of the adamantyl cation has been examined by Polfer et al. (2004).

A systematic study of infrared spectra on different diamondoids—adamantane, diamantane, triamantane, [121]tetramantane, [1(2,3)4]pentamantane, and [12312]hexamantane—was conducted by Pirali et al. (2007). In addition to experimental measurements and owing to more performant computing power, in recent years the details of the vibrational spectra were also investigated via numerical simulations (Banerjee and Saalfrank, 2014).

Figure 3.22 displays experimentally acquired spectra of various diamondoids ranging from adamantane to [12312]hexamantane.

As can be seen from the graphs in Fig. 3.22, the number of spectral features depends on the symmetry of the diamondoid investigated. Higher symmetry leads to less peaks.

In the case of enantiomers, unless a polarized light source is used, there is no difference in the spectra.

3.3.1.2 Raman spectroscopy of diamondoids

Before discussing the main characteristics of Raman spectra of diamondoids in detail, we briefly give an overview of Raman spectra of other carbon nanomaterials, that is, such that are sp^1, sp^2, or sp^3 hybridized. Figure 3.23 displays the Raman spectra of diamond, graphite, the fullerene C60, carbon nanotubes (CNTs), and adamantane.

A first detailed study of the vibrational modes of admantane has been conducted by R. T. Bailey (Bailey, 1971), and later studies on the lower diamondoids up to triamantane also investigated the effect of temperature on the vibrational features (Jenkins and Lewis, 1980). As has been mentioned in Section 3.1.1, the lower

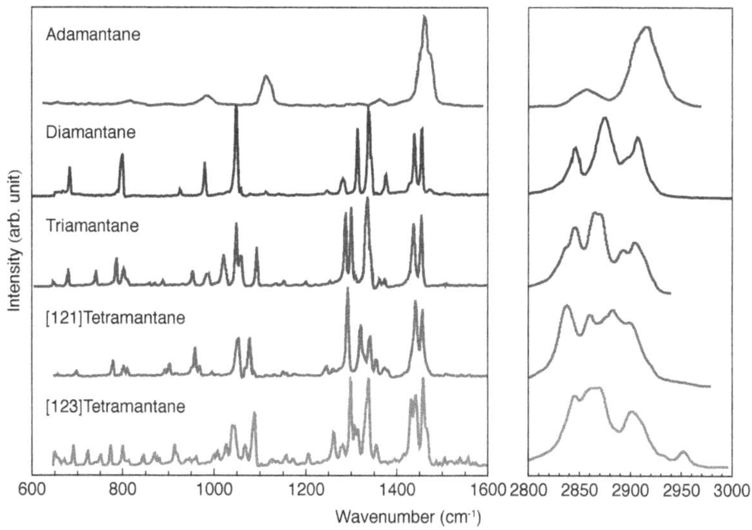

Figure 3.22 Attenuated total reflection Fourier transform infrared (ATR-FTIR) spectra of diamondoids. The data is the one of adamantane (adapted from Garcia et al., 2010c), diamantane, triamantane, and [121]- and [123]tetramantane (data adapted from Oomens et al., 2006). The light source for acquiring the spectra was not polarized; therefore the spectra of [123](P) and (M) tetramantane are the same. The spectrum for tetramantane corresponds to a racemic mixture of the (P) and (M) forms (i.e., equal proportions of the right [dextro-] and left-rotatory [levo-] forms).

diamondoids—adamantane, diamantane, and triamantane—exhibit a transition to a premelting plastic phase. In the case of adamantane, this transition occurs at 209 K, at 447 K for diamantane and at $T > 428$ K for triamantane.

In the following, we briefly discuss the different spectral features of adamantane in more detail. Since adamantane consists of 24 atoms, it has $3N - 6 = 72$ vibrational modes $(\Gamma_{\text{vib}})_{\text{mol}}$,

$$(\Gamma_{\text{vib}})_{\text{mol}} = 5A_1 + A_2 + 6E + 6T_1 + 10T_2 \tag{3.8}$$

with A_i symmetric modes.

As already mentioned, because of the complexity of Raman spectra of diamondoids, numerical simulations can help to interpret such spectra. One example was the theoretical confirmation of

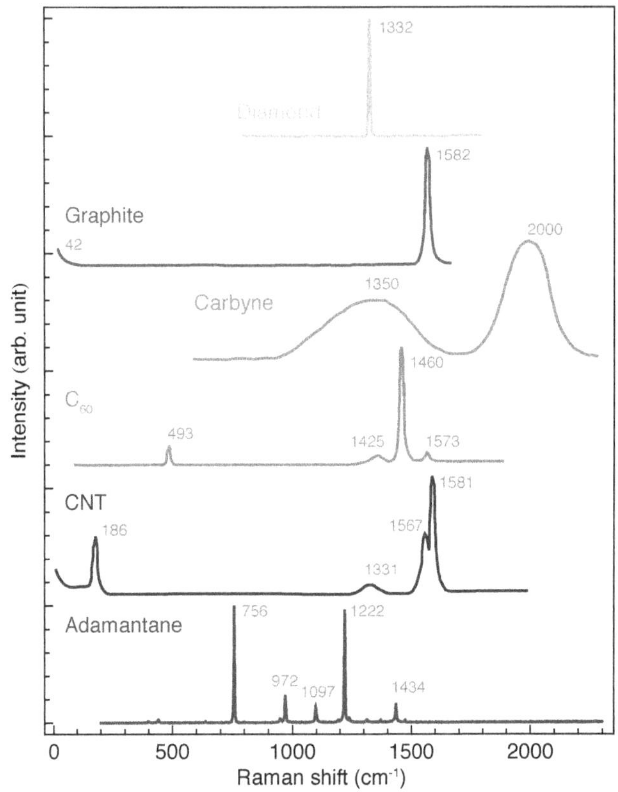

Figure 3.23 Raman spectra of various carbon materials. The data is representative of different carbon materials and hybridizations: diamond (sp^3), graphite (sp^2), carbyne (sp^1), the fullerene C_{60} (sp^2), carbon nanotubes (CNTs, sp^2), and the first diamondoid, adamantane (sp^3).

the experimental Raman spectra of cyclohexamantane (Richardson et al., 2005).

For nonlinear molecules with N atoms, the total number of normal vibrational modes (i.e., modes that are relatively simple such as stretching mode) is $3N-6$ while for linear molecules, it is $3N-5$).

Diamondoid molecules contain specific vibrational modes that can be probed by Raman spectroscopy.

Such symmetry considerations allow to determine the number of theoretically Raman active vibrational modes. Detailed character-

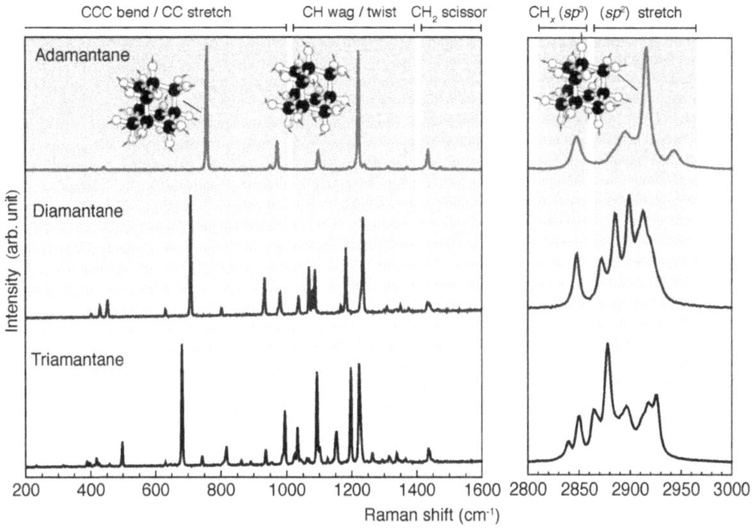

Figure 3.24 Experimental Raman spectra of the lower diamondoids adamantane, diamantane, and triamantane. For easier presentation, the high-frequency domain and the low-frequency domains have been separated. Data adapted from Filik et al. (2006a).

ization of diamondoids using different Raman wavelengths has been conducted by (Filik et al., 2006a).

A series of characteristic Raman spectra of lower diamondoids are shown in Fig. 3.24.

The arrows in the molecule indicate the vibrational modes of the diamondoids.

As in the infrared spectra, Raman spectra of diamondoids can be divided into several regions that are characteristic of certain vibrational modes of the diamondoids.

In addition to the experimental Raman spectra shown in Figs. 3.24–3.28, attempts have been undertaken to simulate these spectra numerically (Filik et al., 2006a,b). In addition, the Raman spectra of cyclohexamantane have been investigated using density functional theory (Richardson et al., 2006).

With increasing cage number n, the vibrational features in both the Raman and infrared spectra become more and more complex. Therefore, to assign the peaks in the experimentally acquired spectra, ab initio simulations have been performed in parallel.

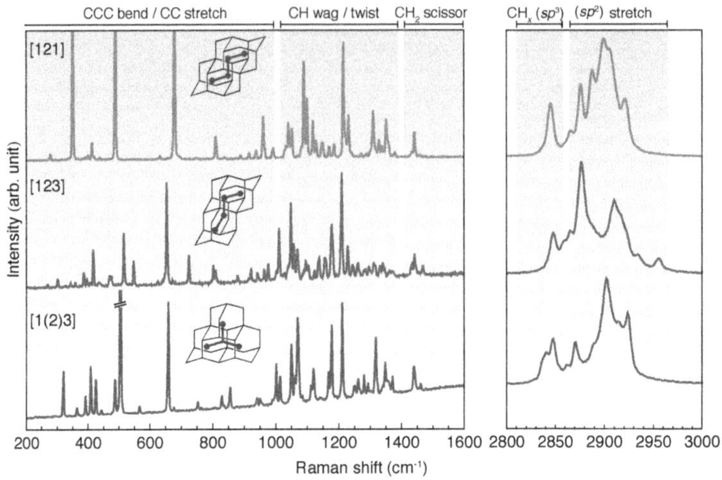

Figure 3.25 Experimental Raman spectra of [121], [123], and [1(2)3]tetramantane. The spectra in the region of 200–1800 cm^{-1} have been normalized with respect to the peak intensity at ~1200 cm^{-1}. Data adapted from Filik et al. (2006a).

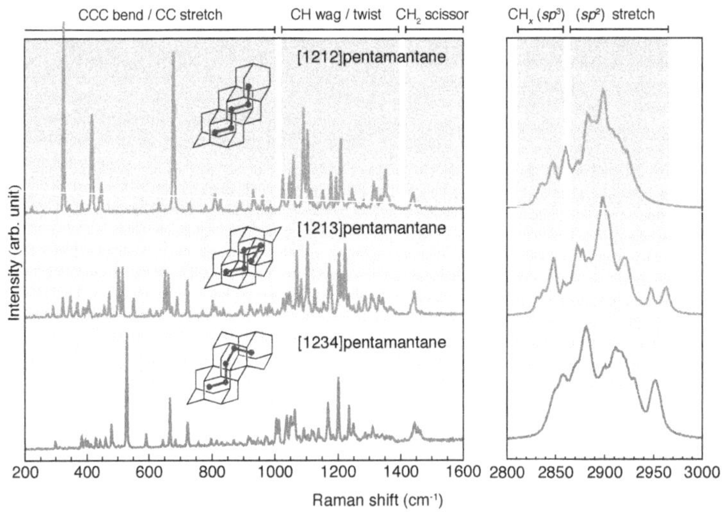

Figure 3.26 Experimental Raman spectra of [1212], [1213], and [1234] pentamantane. The spectra in the region of 200–1800 cm^{-1} have been normalized with respect to the peak intensity at ~1200 cm^{-1}. Data adapted from Filik et al. (2006a).

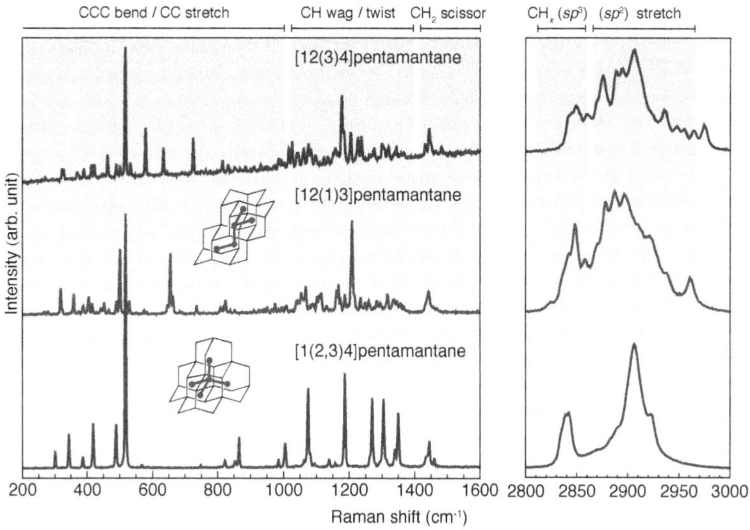

Figure 3.27 Experimental Raman spectra of [12(3)4], [12(1)3], and [1(2,3)4]pentamantane. The spectra in the region of 200–1800 cm^{-1} have been normalized with respect to the peak intensity at ~1200 cm^{-1}. Data adapted from Filik et al. (2006a).

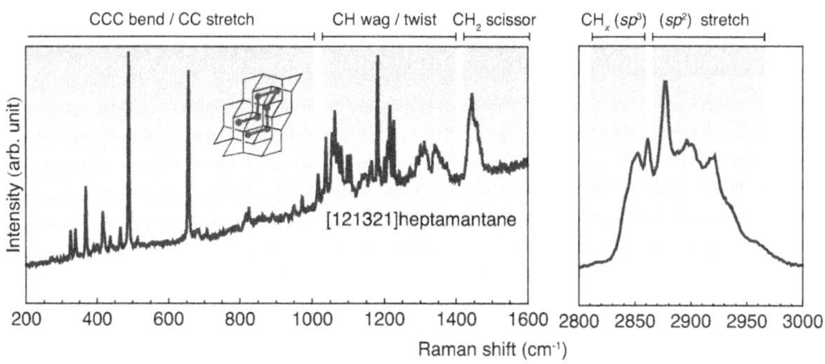

Figure 3.28 Experimental Raman spectra of [121321]heptamantane. The spectra in the region of 200–1800 cm^{-1} have been normalized with respect to the peak intensity at ~1200 cm^{-1}. Data adapted from Filik et al. (2006a).

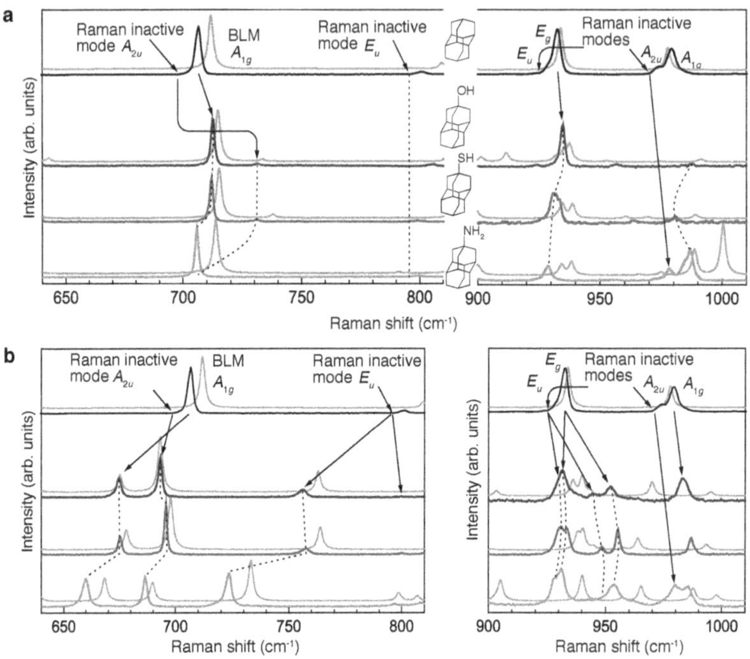

Figure 3.29 Experimental Raman spectra of functionalized diamantane. Data adapted from Meinke et al. (2013).

In the last part of this section, we will have a look at the effect of the functionalization on the modification of Raman spectra.

However, while such Raman spectra allow to understand the modifications, currently it is still not possible to simulate the experimental spectra with 100% accuracy. One of the reasons is that the ab initio simulations of Raman spectra involved the calculation of the derivative of the polarization tensor, which is computationally time consuming.

In addition, in the simulations, only isolated molecules are considered, whereas in real Raman spectra, the nature of the crystal lattice leads to additional peaks and shifts. Table 3.2 lists vibrational modes of adamantane ([0]), diamantane ([1]), and select triamantanes and pentamantanes.

Raman scattering experiments of triamantane in the low frequency range (~0–100 cm^{-1}) for temperatures between 40 and

Table 3.2 Summary of known vibrational modes of diamondoids. We list again the symmetry of the molecules. Vibrational modes that are not Raman active are marked with the superscript^{na}, and degenerative modes with^{g}. Data adapted from Filik et al. (2006a)

		Vibrational modes																	Raman
Diamondoids	Symmetry	Total	A_1	A_2	A_{1g}	A_{1u}	A_{2g}	A_{2u}	B_1	B_2		B_g	B_u	E	E_g	E_u	T_1	T_2	active
[0]		72	5na	1							6			7na	11	**22**			
[1]		96			11	6na	5na	10na				16	16na			**27**			
[12]	C_{2v}	120	35	25					29	31						**120**			
[121]	[1(2)3]	C_{3v}	144	29	19na							48^{d}					**77**		
[1212]																			
[1(2, 3)4]	T_d	168	10	4na							14			18na	24	**48**			

340 K showed a phase transition at about 285 K (Jenkins, 1978). This phase transition was found to be consistent with previous XRD measurements and are due the change from an ordered structure (D_{2h}^{24}, $Z = 16$) to a disordered structure (D_{2h}^{25}, $Z = 2$).

Recently, the effect of pressure on Raman and XRD spectra in [121]tetramantane has also been investigated (Yang et al., 2014).

Functionalizing diamondoids leads also to changes in both their infrared and Raman spectra. For example, both Raman and infrared spectra of functionalized diamondoids have been analyzed both experimentally and theoretically using density functional theory analysis (Bistricic et al., 2002), see Fig. 3.30, which shows the

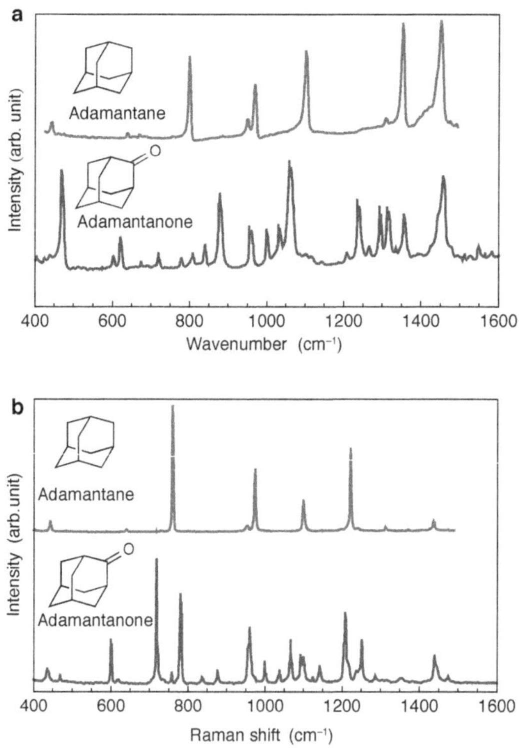

Figure 3.30 Infrared and Raman spectra of adamantane and adamantanone. (a) Infrared spectra of adamantane and adamantanone. (b) Raman spectra of adamantane and adamantanone. Data adapted from Bistricic et al. (2002).

experimental infrared and Raman spectra of adamantane and adamantanone.

In short, vibrational spectroscopy can be used for "fingerprinting" diamondoids. However, for this, it is necessary to have sufficiently quantities of purified material at hand, typically on the order of milligrams, otherwise it is not possible to conduct the spectroscopy measurements.

3.4 Mass Spectrometry of Diamondoids

While vibrational spectroscopy methods are relatively easy to conduct, the requisite is that one has a purified sample, preferably in crystalline form. In addition, these methods are only applicable if the products are available in at least milligram quantities.

As will be discussed in later chapters, with the exception of lower diamondoids and higher diamondoids extracted and isolated from oil and gas reservoirs, they are not available in large quantities.

Another method that allows analyzing products is gas chromatography–mass spectrometry (GC-MS), which enables the detection of samples of very small quantities.

In this section, we will give an overview of mass spectra of diamondoids and their main features.

The mass spectra contain usually a prominent molecular ion peak, that is attributed to the molecular ion (denoted $M^{+\bullet}$). Since diamondoids are closed-shell molecules, singly ionized molecules are odd-electron ions (i.e., open shell). In addition, depending on the ionization energy, the mass spectra often contain a peak at M/2, which is due to the double ionized molecule, M^{2+}.

Figure 3.31 displays the mass spectra of adamantane and diamantane, the data being used from the NIST Mass Spectral Library (NIST/EPA/NIH Mass Spectral Library (EI), 2012).

3.4.1 Nuclear Magnetic Resonance Spectroscopy of Diamondoids

While vibrational spectrosocopy is the method of choice for characterizing diamondoids, to obtain meaningful data, it is necessary

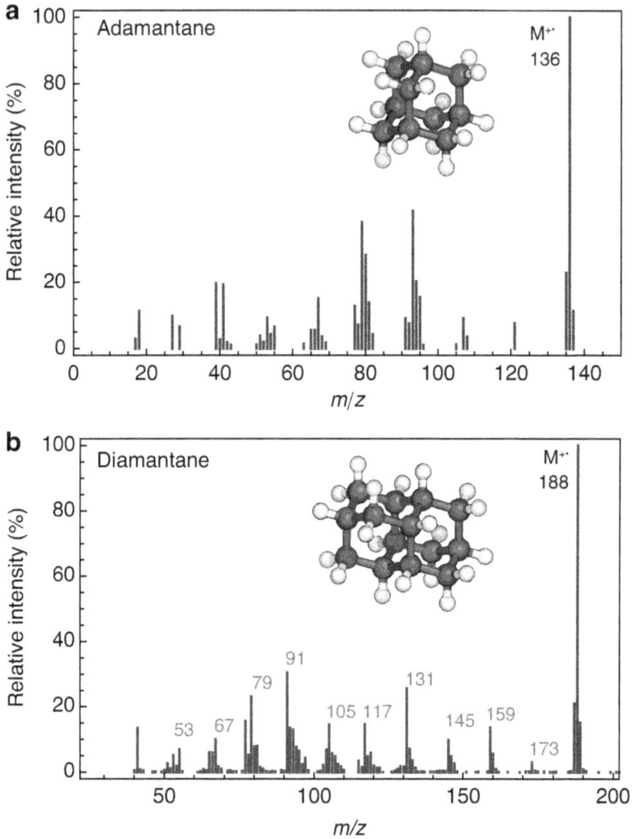

Figure 3.31 Mass spectra of adamantane and diamantane. (a) Adamantane and (b) diamantane. The peaks at m/z 136 and 188 are attributed to the respective molecular ions $M^{+\bullet}$ of adamantane and diamantane, while the others correspond to fragment ions. The peaks at M/2 are due to the doubly ionized molecule M^{2+}.

to have sufficient quantities of diamondoids (typically on the order of milligrams). As has been shown in the previous sections, the vibrational spectra of especially the lower diamondoids are well known and with the aid of computational simulations, also allow to assign peaks of higher diamondoids (Filik et al., 2006a). However, in cases where the exact substance is not known, it is necessary to resolve to other techniques.

Table 3.3 ^{13}C chemical shifts for adamantane and diamantane

Name	Molecular formula	Dualist code	Crystal structure
Adamantane	$C_{10}H_{16}$	[0]	**Fm$\overline{3}$m** (O_h)
Diamantane	$C_{14}H_{20}$	[1]	**Pa$\overline{3}$** (T_h^6)

While not as widely employed as vibrational spectroscopy methods, nuclear magnetic resonance (NMR) techniques have also been used to characterize diamondoids, but so far, they include mainly the smallest diamondoids, that is, adamantane, diamantane, and certain derivatives (Hajek et al., 1985; Horska et al., 1988). One exception is the work of Panich et al. (2006), who investigated NMR spectra of ultrananocrystalline diamonds.

Table 3.3 lists the ^{13}C chemical shifts of adamantane and diamantane.

Chapter 4

Current and Future Applications of Diamondoids and Their Derivatives

In this chapter, we will briefly present a few of the current and possible future applications of diamondoids and their derivatives. Because of their advantageous properties, especially with respect to thermal and chemical stability, they hold promise for a wide range of possible applications. The first part of this chapter focusses on diamondoids used as probes in oil and gas reservoir exploration, while the second treats current advances in the use of diamondoids in chemistry, pharmaceutics, medicine, and biotechnology. The third part presents examples of the use of diamondoids in materials science and nanotechnology, while the last part gives a brief outlook on possible future applications.

4.1 Overview

In recent years, diamondoids have elicited increasingly interest as possible nanomaterials. Work by several groups has suggested that their special properties—high thermal stability, well-defined structure, and no known toxicity—makes them interesting for

Diamondoids: Synthesis, Properties, and Applications
Sven Strauss and Kazuo Terashima

ISBN 978-981-4745-18-5 (Hardcover), 978-1-315-36466-7 (eBook)
www.panstanford.com

applications for a wide range of applications (Freitas and Merkle, 2008).

These possible applications of diamondoids include pharmaceutics and chemistry, biotechnology, and nano- and optoelectronics. However, to date, one of the most common applications of diamondoids is as markers for the evaluation of the maturity of petroleum and gas fields.

In the following sections, we will briefly present these applications more in detail, together with relevant works from the literature. We have divided these different applications into different fields, but these have been chosen more or less arbitrarily. At the time of writing of this book, we tried to include the latest research on diamondoids; however, since this is still a work in progress, new applications are certain to be found.

4.2 Applications of Diamondoids in Oil Exploration

Owing to their high thermal stability compared to alkanes of similar molecular weight, diamondoids persist in petroleum reservoirs, leading to a gradual increase in their concentration during cracking (Dahl et al., 1999). This diamondoid concentration can be used to identify the possible occurrence and extent of oil destruction, and the oil deadline in a particular basin. The oil deadline describes the conditions where oil can be formed: typically at depths of 5 km and temperatures of 150°C–175°C (recent experiments show that petroleum can persist at temperatures higher than 200°C). Otherwise, there is onset of oil cracking, that is, formation of gas and pyrobitumen.

Because of their high thermal stability, diamondoids find application in characterizing natural gas condensates and gasoline (Stout and Douglas, 2004). Another important application of diamondoids is in environmental fingerprinting and environmental forensics. The relative abundance of different diamondoids or their derivatives, especially those of adamantane and diamantane, and the high heat resistance can be used. Consequently, diamondoids have been used to characterize oil and gas reservoirs in many different locations. For example, diamondoids were used to assess the maturity of gas condensates in the Southern Indus Basin (Pakistan) (Nasir and Fazeelat, 2013). Moreover, the presence of diamondoids allows for identifying the source in oil spills (Lengger et al., 2013; Mansuy et al., 1997).

4.2.1 Formation of Diamondoids in Natural Gas Reservoirs

Diamondoids found in petroleum sources result from carbocation rearrangements of organic precursors, for example multiringed terpene hydrocarbons deposited on clay mineral superacids. It has been suggested that during oil cracking, they are neither formed nor destroyed but conserved and their increased concentration can be used for estimating the extent of conversion of liquid to gas and pyrobitumen. The formation of lower diamondoids is mainly attributed to (1) Lewis acid–catalyzed rearrangements of polycyclic hydrocarbons (Lin and Wilk, 1995; Wingert, 1992); (2) high-temperature cracking of high-molecular-mass fractions (Giruts et al., 2006; Giruts and Gordadze, 2007).

Figure 4.1a illustrates the principle of a Lewis acid and base: A Lewis base provides two electrons, whereas a Lewis acid is an electron pair acceptor. The reaction in Fig. 4.1b shows the

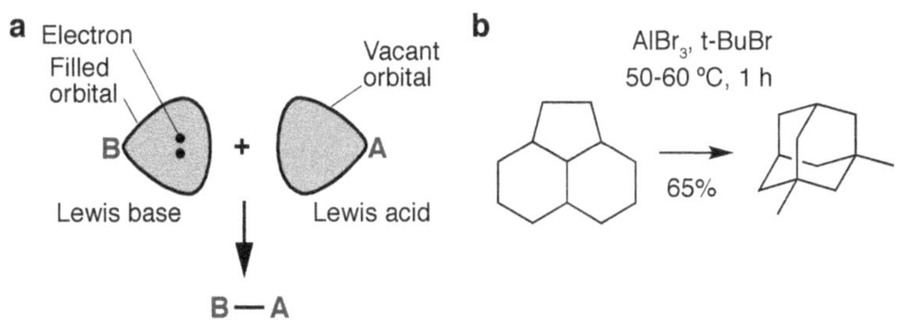

Figure 4.1 Example of Lewis acid catalyst. (a) Lewis acid: acceptor of two electrons. (b) Example of a Lewis acid–catalyzed rearrangement of a polycyclic hydrocarbon: formation of 3,5-dimethyladamantane from perhydrogenated phenanthrene in $AlBr_3$ (aluminum bromide) and *t*-BuBr (*tert*-butyl bromide).

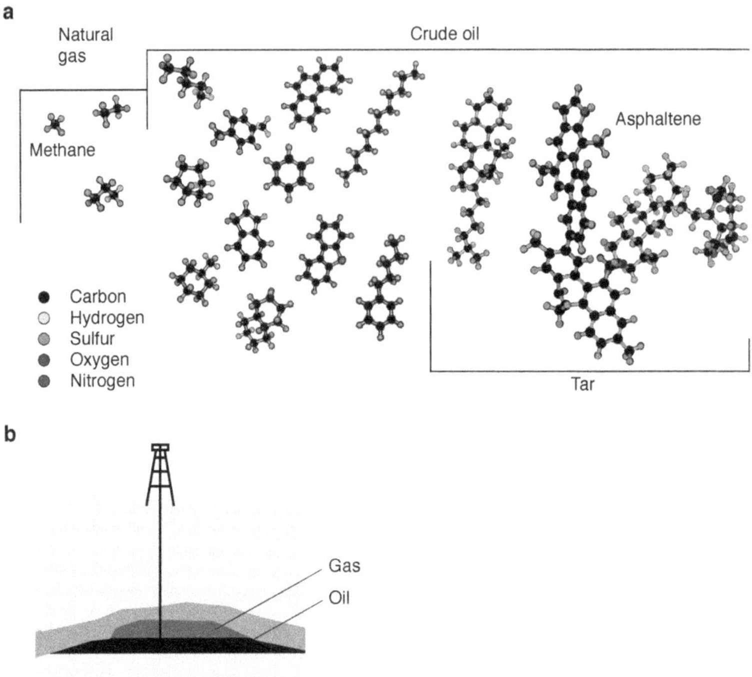

Figure 4.2 Overview of possible natural sources for the formation of lower diamondoids. (a) Natural gas and crude oil. (b) Oil reservoir.

catalyzed rearrangement of a polycyclic hydrocarbon, perhydrogenated phenanthrene, to 3,5-dimethyladamantane in aluminum bromide.

Laboratory experiments have shown that lower diamondoids (i.e., $n \leq 4$) can be obtained from different natural sources: sedimentary rocks, peats, kerogens, and crude oil and their different group components (i.e., saturated and polar fractions [resin and asphaltene]) and, finally, carbon compounds (C_{16}, C_{19}, C_{22}, C_{34}, C_{36}, and *n*-alkanes).

Figure 4.2 gives an overview of different natural sources of diamondoids.

Despite their high thermal stability—especially when compared to other carbon materials such as alkanes, etc.—studies have shown that lower diamondoids can be destroyed at temperatures that are higher than those during their generation (Fang et al., 2012; Wei et al., 2006), that is, at temperatures above $T = 340°C$.

At higher maturation levels, C_3- and C_4-alkylated adamantanes are probably more stable than other adamantanes.

The formation of lower diamondoids in petroleum consists of three main stages: (1) early generation during the formation of oil (~0.8–1.0% EasyRo), (2) generation during the cracking of oil (>1% EasyRo), and (3) destruction of diamondoids during the late stages of oil cracking ($\gtrsim$2% EasyRo).

The most abundant adamantane and diamantane sources are saturated oil fractions in the maturity range of 1.0%–2.3% for adamantane and 1.6%–2.5% for diamantane. The second-most abundant source of adamantanes and diamantanes is resin and aromatic fractions. In contrast, the contents of diamondoids in the asphaltene fraction was almost negligible. Finally, in the evolution curve of diamondoids, rapid destruction is observed.

The origins and formation mechanisms of diamondoids in oils and source rocks are still a matter of debate and many studies being conducted in order to understand the conditions (i.e., temperatures and pressures, but also the type of rocks and sediments) that can lead to the formation of diamondoids (Fang et al., 2013, 2012).

Diamondoids are absent in recent organic matter; therefore they are probably not of biosynthetic origin, that is, formed within living organisms (bacteria, etc.) or cells. Currently, diamondoids are thought to be formed during diagenesis (i.e., the formation of sediments to sedimentary rocks) and the catagenesis (i.e., the cracking process during which organic kerogens are converted into hydrocarbons) of organic matter. Kerogen is a collective name for an organic material that is insoluble in organic solvents, water, or oxidizing agents. Kerogen consists of very large molecules (and could be considered a polymer). When exposed for sufficiently long times and high temperatures, kerogens again crack down to smaller molecules, mostly petroleum.

In addition to natural oil and gas reservoirs, diamondoids, nanodiamonds, and diamondoids have also been observed in space (Lewis et al., 1987).

4.3 Current and Possible Future Applications of Diamondoids and Derivatives in Chemistry, Pharmaceutics, Medicine, and Biotechnology

4.3.1 Applications in Chemistry

4.3.1.1 Host–guest chemistry

Host–guest chemistry involves binding of a guest molecule by a host molecule, the bond between the molecules being noncovalent. Noncovalent bonds can involve hydrogen bonds, ionic bonds, van der Waals forces, or hydrophobic interactions. Examples of diamondoids acting as hosts in host–guest chemistry are given in (Voskuhl et al., 2012). In this specific work, diamondoids are included inside sugar rings. There are other possible guest molecules, for example, cyclodextrins, calixarenes, pillararenes, cucurbiturils, porphyrins, metallacrowns, crown ethers, zeolites, cyclotriveratrylenes, cryptophanes, carcerands, and foldamers. One example of a host molecule that is obtained by substitution by 1-adamantyl is calix[4]arene.

Diamondoids—specifically adamantane—have also been used as a seed for diamond film growth (Tiwari et al., 2010, 2011).

Figure 4.3 gives an example of a host–guest chemistry based on adamantane. In this particular case, methyladamantane units are attached to cyclodextrin.

4.3.2 Applications of Diamondoids in Pharmaceutics and Medicine

Polycyclic cage compounds (Marchand, 1995), to which diamondoids belong, have important pharmaceutical applications: They have shown potential for the treatment of neurodegenerative diseases such as Parkinson's and Alzheimer's diseases, and they also possess activity against virus infections such as influenza virus A, human immunodeficiency virus (HIV), and others. The potential of diamondoids as anti-Parkinson's agents was suggested already in the 1970s (Chakrabarti et al., 1976). The chemical stability and well-defined cage structure of diamondoids are the reason that they

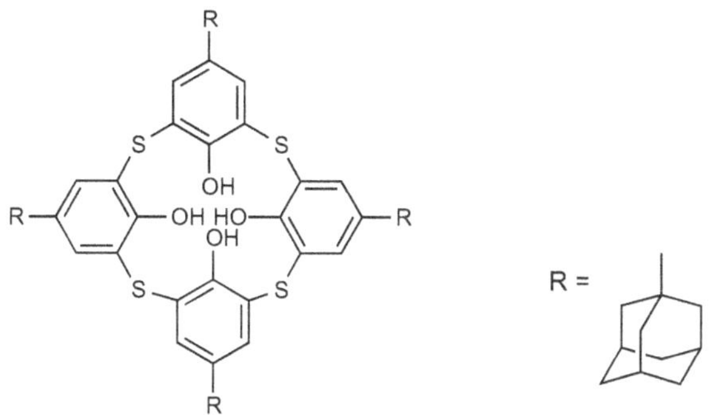

Figure 4.3 Illustration of host–guest chemistry based on adamantane. The methyladamantane units, indicated by R in the cyclodextrin molecule, are attached to the aromatic rings in the structure.

can act as useful scaffolds for the attachment of side chains, which permits the synthesis of drugs. In the following section, we will discuss the properties of the most commonly used adamantane derivatives that so far have been used for the treatment of various diseases.

The first adamantane derivative that was discovered to show promise for pharmaceutical applications was adamantane amine (IUPAC name 1-amino-adamantane) (see Fig. 4.4), also commonly named amantadine, which showed activity against different viruses (influenza, hepatitis C, and herpes zoster neuralgia). Amantadine has been found to inhibit the decapsidation stage of influenza virii,

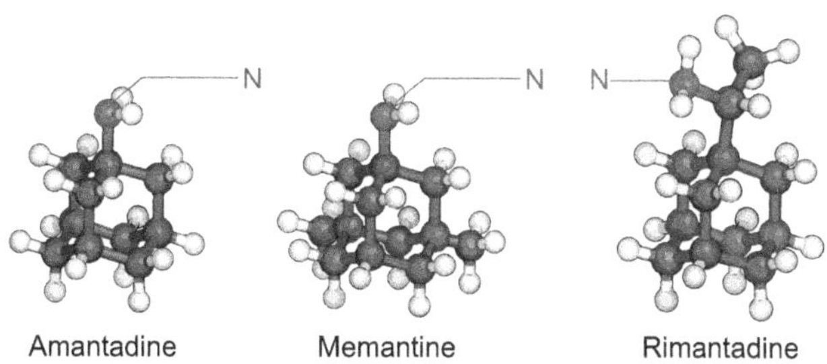

Figure 4.4 Molecular structures of the currently most commonly used adamantane derivatives in pharmaceutics: amantadine, memantine, and rimantadine.

that is, the removal of the protein shell (capsid) that encloses the genetic material of a virus, thereby obstructing its reproduction. Another function amantadine is believed to perform is to effectively block ion channels for virus entry, by forming hydrogen bonds with a histidine residue (His-37), in the M2 transmembrane domain protein of the influenza virus.

Rimantadine (alpha-methyl-1-adamantane-methylamine hydrochloride) is another adamantane derivative, in addition to amantadine and memantine (see Fig. 4.4). Amantadine (1-adamantanemine hydrochloride) and memantine (displayed in the middle of Fig. 4.4) both show anti-Parkinsonian activity. In the following paragraphs, we treat the case of rimantadine. In 1993, rimantadine, which is the alpha-methyl derivative of amantadine, was approved by the US Food and Drug Administration for the prophylaxis and treatment of influenza A virus (although later studies suggest that the activity against influenza A decreased). Compared to amantadine, rimantadine has a few advantageous properties:

(1) It has fewer central nervous and sleep side effects.
(2) Its elimination half-life, that is, the time it takes the body to reduce its quantity by half, is about 24 to 36 hours, compared to 12–18 hours for amantadine.
(3) It possesses a 10 times higher partition coefficient when compared to amantadine.

With regard to the first point, a study on 251 healthy adults (Evidente et al., 1999) showed that daily doses of 300 mg.day^{-1} of amantadine lead to significant central nervous system side effects. These included nervousness, lightheadedness, and concentration troubles. The subjects were also four times more likely to experience sleep disturbance, namely insomnia or fatigue.

Concerning the second point mentioned above, a longer half-life means that a single daily dose is sufficient for treatment. And a higher partition coefficient, which expresses the ratio of equilibrium drug concentration in lipids or fatty acids to that in water, leads to larger intracellular concentrations and faster transport across the blood–brain barrier of rimantadine. Studies conducted using animal models showed that rimantadine enters the brain 10 times more rapidly than amantadine.

In the treatment of Parkinson's disease patients, about 80% of patients reported a self-perceived improvement (Evidente et al., 1999). However, there are also a few disadvantages related to the use of rimantadine. The first is that it loses its effect against Parkinson's disease after several months, similar to amantadine. In addition, for some patients, it can lead to other side effects.

The exact molecular mechanism of how rimantadine leads to an improvement in Parkinson's disease patients is still not entirely clear. Studies suggest that its possible mechanisms are similar to that of amantadine, whose anti-Parkinson's effects include increased synthesis, release, or diminished reuptake of dopamine, a neurotransmitter whose quantity is diminished in Parkinson patients, reducing their motor ability. Furthermore, amantadine also leads to anticholinergic effects, that is, it inhibits the physiological action of acetylcholine, a neurotransmitter, thereby blocking involuntary muscle movements caused by diseases such as asthma, incontinence, gastrointestinal cramps, and muscle cramps.

Alzheimer's disease has become one of the most common diseases of aged populations, especially in industrialized countries. Estimates of affected people are in the range of 15 million people, and the number is constantly rising (Reisberg et al., 2003). Alzheimer's disease results in reduced cognitive and motor capacities in affected persons and in moderate-to-severe stages impairs significantly not only mental capabilities but also instrumental and daily basic activities. While there are pharmaceutical drugs for treating mild-to-moderate Alzheimer's disease, currently there are no treatments for advanced stages. One reason for the reduced cognitive capabilities as Alzheimer's disease progresses is the overstimulation of a specific receptor, *N*-methlyl-D-aspartate (NMDA), by glutamate, one of the main neurotransmitters in the brain. In addition to Alzheimer's disease the NMDA receptor is also implicated in other brain-related diseases, reduced memory capabilities, and dementia. A 28-week study in 252 moderate-to-severe Alzheimer patients aged at least 50 years old and more than 75 years on average showed significant reduced degeneration on different impairment scores when being administered daily doses of 20 mg of memantine (1-amino-3,5-dimethyladamantane), an adamantine derivative. That study indicated that memantine acts as an NMDA antagonist, reducing the effect of overstimulation by glutamate, thereby alleviating symptoms related to Alzheimer's disease. In 2003, memantine was approved by the Federal Drug Administration (FDA) under the trade name Namenda® for the treatment of Alzheimer patients.

In addition to the three main adamantane derivatives—amantadine, memantine, and rimantine—discussed above and that have been found to show promise as pharmaceutical drugs, following the discovery of the positive effects of amantadine in treating different diseases, other adamantane derivative–based compounds have been investigated over the years.

The first orally active antiviral drug used in the United States was amantantadine hydrochloride. The beneficial effect of amantadine on Parkinson's disease was discovered accidentally when patients were administered adamantine for treating influenza infections. Studies indicate that one of the reasons adamantine leads to improvement in the condition of Parkinson patients by either increasing dopamine levels by inhibiting dopamine reuptake or stimulating dopamine release. Furthermore, by docking to binding sites located in receptor/ion-channel complexes, adamantine blocks the uptake of calcium ions into neurons. Over the years, various types of adamantane derivatives have been synthesized for use in pharmaceutics. The advantage of using adamantane is that it can both act as a scaffold for functional groups, especially side chains, and, therefore, improve a drugs lipophilicity, that is, its capability to be dissolved in lipids or fats. This means that a drug can be transported more easily across cellular membranes, and it may lead to easier attachment of the drug in receptors. Finally, such adamantane-based drugs afford metabolic stability, which leads to a prolonged effect of the drug. This means that both the drugs' dosing quantity and frequency can be reduced, helping to improve patients' compatibility and compliance.

Figures 4.5 and 4.6 show the molecular structures of adamantane derivatives that have been used for various pharmaceutical applications.

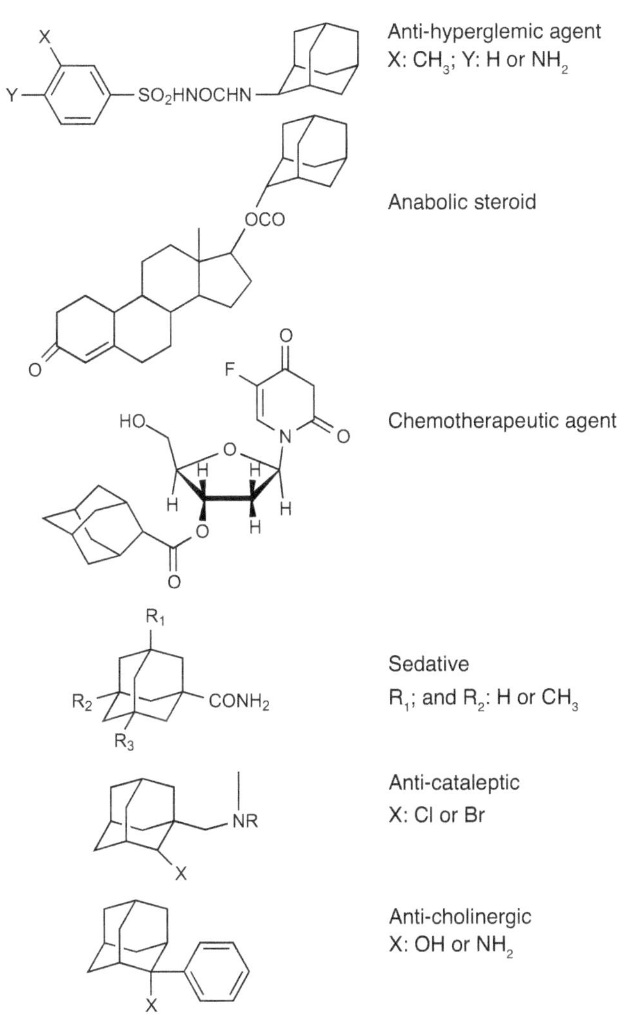

Figure 4.5 Molecular structures of adamantane derivatives and their activity against specific diseases.

In summary, adamantane derivatives have shown promise for a wide range of drugs and for treating various diseases and infections. While currently, there are no drugs based on diamondoids other than adamantane, it is expected that because larger diamondoids

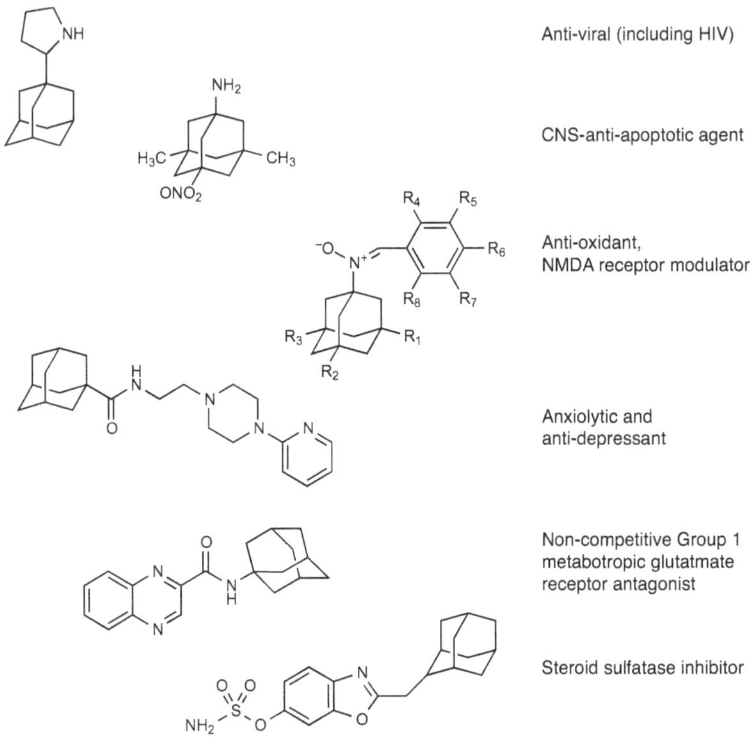

Figure 4.6 Molecular structures of adamantane derivatives and their activity against specific diseases (continued).

have a similar molecular structure as adamantane, they could also be useful for pharmaceutical applications.

4.3.3 Current and Possible Future Applications of Diamondoids for Drug Delivery

Drug delivery is the attachment of a given drug to a carrier, called vector, which transports the drug to the destined location in the cell and then releases it. As already mentioned previously, one of the attractive properties of diamondoids, in particular adamantane and its derivatives, is their lipophilicity. This means that they can be easily be taken up by lipidic membranes in cells. Consequently, in the case of drugs that are not readily absorbed by cells, when attached to diamondoid units, their incorporation in the organism can be increased.

Because of the well-defined structure and surface chemistry of diamondoids, they are attractive as bonding sites of functional groups. When adamantane or other diamondoids are functionalized by short peptide sequences, these can act as binding sites for different types of macromolecules: proteins, nucleic acids, lipids, and polysaccharides, to name a few examples.

Several examples of the application and potential of adamantane derivatives acting as vectors for drug delivery have also been demonstrated. One was the increased uptake of various azidothymidine (AZT) drugs attached via an ester spacer to 1-adamantyl (Tsuzuki et al., 1994). Compared to the same drug without the 1-adamantyl as vector, its uptake in the brain tissue was increased by a factor varying from 7 to 18.

In recent years, it has been found that commercial nanodiamond particles produced by detonation synthesis and possessing diameters of approximately 5 nm are interesting for biotechnological and medical applications because of their low cell toxicity, biocompatibility, and structural stability. Furthermore, because their surface chemistry can be adjusted, they are attractive as bases for the attachment of functional groups, making them attractive as drug delivery systems.

As has been shown in Section 2.6, the surface chemistry of nanodiamonds can be very diverse and they have to be tailored by different surface treatments such as oxidation. To attach drugs to nanodiamonds, different mechanisms are possible. These can either be covalent bonding or adsorption, the type of bonding often depending on the purpose of the intended application. For drug delivery, adsorption/desorption processes are often preferred because they are simple and they do not involve any change in the molecular structure of the drug that would take place in the case of covalent binding.

Additionally, the adsorption of drugs (especially those that are only poorly soluble) on nanodiamonds offers also the advantage that they can be taken up more readily by the organism. This is in part due to the biocompatibility of nanodiamonds and the higher exposure of the drug to the aqueous environment: A monolayer of drug adsorbed on a nanodiamond particle offers maximal exposure compared to a monocrystal of the same drug.

However, especially for cancer treatment, the binding of the drug to the nanodiamond has to be well controlled in order to avoid overloading of the organism by the drug in the case of strong adsorption, or premature release of the drug that could lead to side effects. Recent studies (Mochalin et al., 2013) have therefore investigated in detail the binding strength of drug molecules to nanodiamonds, in addition to desorption heat and kinetics. Other factors that have to be taken into account for the application of nanodiamonds as drug delivery systems are the role of pH, the ionic composition of the fluid, nanodiamond purity and dispersion, and the surface chemistry.

While nanodiamonds have been applied for various applications in medicine and biotechnology, for example, blood cleansing and sorption of heavy metals, dyes, proteins, toxins, and viruses, only little is known about the fundamental adsorption/desorption thermodynamics and kinetics of nanodiamonds.

A recent study showed that the adsorption of drugs strongly depends on their surface chemistry and purity.

Currently, there are no studies available on higher diamondoids, similar in size to nanodiamonds, that could have acted as drug delivery systems, it is surmised that because of the well-defined molecular structure and surface terminations, diamondoids could alleviate some of the difficulties encountered in the use of nanodiamonds, especially control of the surface chemistry and size distribution.

4.4 Applications in Materials Science and Nanotechnology

4.4.1 Materials Science

Because of their high thermal stability, diamondoid derivatives are used increasingly as additives in various materials to improve their properties. For example, Idemitsu Chemicals Co. is selling various adamantane derivatives under the generic name Adamantate^{TM}.

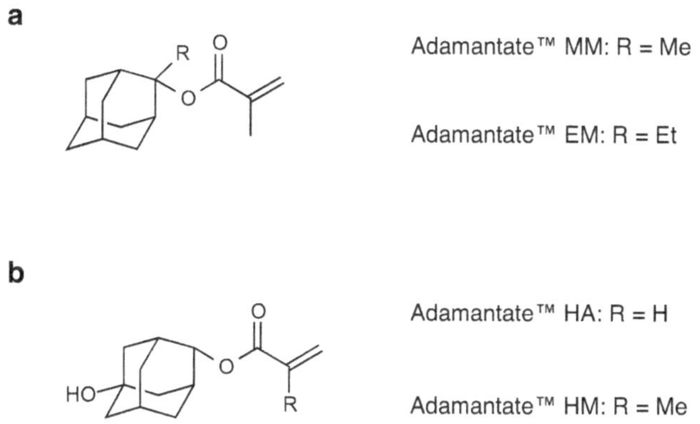

Figure 4.7 Examples of adamantane derivates commercialized under the name Adamantate™ and used in various application ranging from panel displays to photoresists. (a) Adamantate™ MM and EM. (b) Adamantate™ HA and HM. The abbreviation Me stands for methanol and Et for ethanol.

Using Adamantate™ as a base resin, superior characteristics such as increased heat resistance, stiffness, and transparency can be added to a variety of products including flat-panel displays and photoresists used in semiconductor manufacturing (cf. www.idemitsu.com/products/petrochemicals/chemicals/adamantate.html). Figure 4.7 displays the molecular structure of these adamantane derivatives.

The potential of diamondoids for nanotechnology, especially for electron emission, has been recognized recently (Drummond, 2007). In addition to this, because of the thermal stability, diamondoids have been used to improve thermal resistance of polymers. One example is the development of high-temperature polymers incorporating diamantane (Malik et al., 1991). Another example is the incorporation of diamondoids in photopolymers such as SU-8 to form a nanocomposite polymer (Chiamori et al., 2008).

In addition to adding stability to polymers, diamondoids have also been suggested as subnanosized molecular building blocks. For example, functionalized adamantane was proposed as blocks for building self-assembled nanostructures (Garcia et al., 2009). The same group also carried out numerical simulations to investigate the possibility of boron- and nitrogen-functionalized diamondoids for self-assembly (Garcia et al., 2010b), especially adamantane (Garcia et al., 2010a).

The effect of nanoscale diamondoids on the thermomechanical and morphological behaviors of polypropylene and polycarbonate was studied by Ghosh et al. (Ghosh et al., 2007).

A recent review paper explores the possibility of diamondoids for forming self-assembled monolayers (SAMs) in detail (Hohman et al., 2010). Finally, diamondoids have been suggested as a means to form advanced composites (Hu et al., 2006). Derivatives of adamantane, such as Adamantate^{TM}, are used as additives in photoresists to increase the thermal resistance and by incorporating diamondoids in metallic coatings, the wettability could be increased (Schwertfeger and Schreiner, 2010).

A few of these examples will be examined in more detail in the following sections. While many of these examples are mainly numerical simulations, there are a few cases where the potential of diamondoids for certain applications has also been demonstrated by experiments.

4.4.2 Diamondoids as an Electron Source

Diamondoids could also be used as semiconductors possessing negative electron affinity (NEA). Previously, it has been reported that large-area SAMs of a functionalized diamondoid, [121]tetramantane-6-thiol, could be used for highly monochromatic electron photoemission (Yang et al., 2007). Figure 4.8 shows the molecular structure of [121]tetramantane-6-thiol and a schematic of an SAM, and typical photoemission spectra for [121]tetramantane-6-thiol SAMs on Au and Ag are displayed in Fig. 4.9a and 4.9b, respectively.

Work by Willey et al. (Willey et al., 2008) has shown that the diamondoid monolayer structure and thiol substitution can be controlled, permitting to change the electronic structure of the SAMs.

The mechanism behind the monochromatic photoelectron emission was investigated more in detail by Clay et al. (Clay et al., 2009) and is shown schematically in Fig. 4.10. The authors found that the reason for monochromatic photoelectron emission of diamondoid SAMs is that like diamond, diamondoids possess NEA. It has been

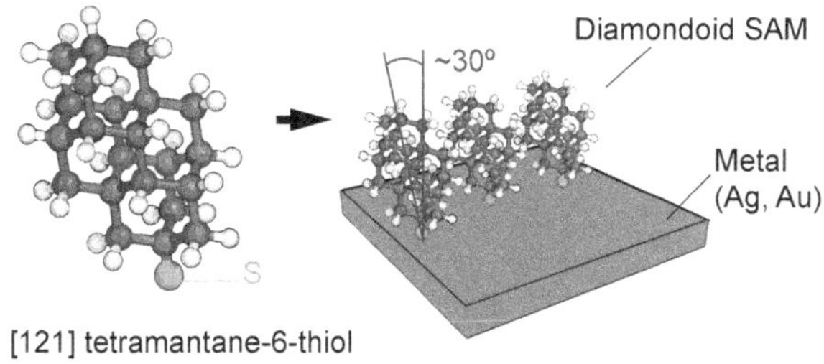

Figure 4.8 Molecular structure of [121]tetramantane-6-thiol and schematic of self-assembled monolayers. Near-edge X-ray absorption fine structure (NEXAFS) measurements revealed that the orientation of the [121]tetramantane-6-thiol molecules, that is, the angle measured between the S—C bond and the surface normal, had values of 30 ± 10°.

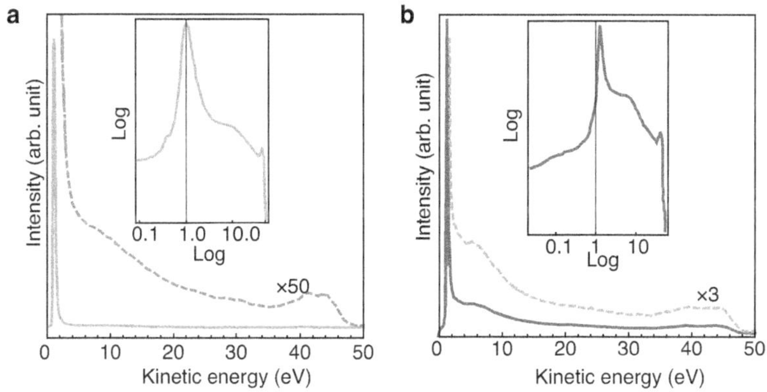

Figure 4.9 Photoelectron emission spectra of [121]tetramantane-6-thiol self-assembled monolayers (SAM). (a) Emission for Ag SAM. (b) Emission for Au SAM. In both graphs, the dotted lines represent the data that has been enlarged by ×50 for Ag and ×3 for Au to enhance the features of the valence bands. Data adapted from Yang et al. (2007).

found that SAMs of [121]tetramantane-6-thiol on Au or Ag films have quantum yields larger than unity.

The NEA of diamondoids results from unoccupied states in the lowest unoccupied molecular orbitals (LUMOs), which, when populated by an electron, directly leads to spontaneous electron emission (Roth et al., 2010). In contrast to Si and Ge nanoparticles,

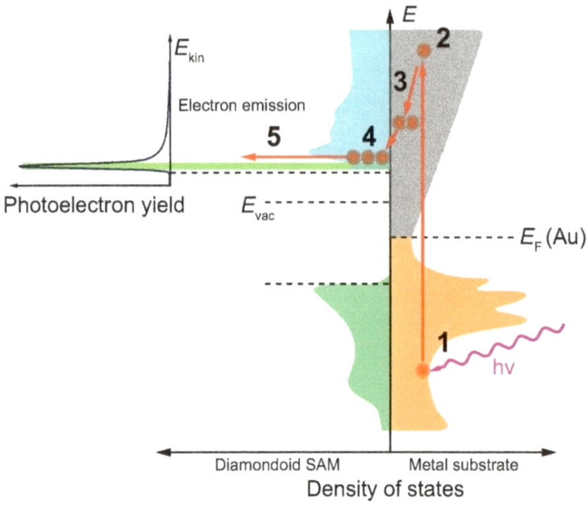

Figure 4.10 Mechanism of photoelectron emission from diamondoid self-assembled monolayers. Schematic of the currently understood processes leading to spontaneous photoelectron emission: **1** electrons are excited by photon and excited to unoccupied states with $E > E_f$, **2**; **3** thermalization of electrons to lower-energy levels; **4** electrons with $E > E_{CBM}$ move to diamondoids; **5** by transitioning to lower-energy levels and accumulating at the bottom of the conduction band, these electrons excite phonons in the diamondoid molecules and, as a consequence of the negative electron affinity, transfer into vacuum. E_f: Fermi level of gold (Au) substrate; E_{vac}: vacuum level; E_{kin}: kinetic energy; E_{CBM}: conduction band minimum. Adapted from Yang et al. (2007).

where the LUMO is core confined, the LUMO of diamondoids is a delocalized surface state (Drummond et al., 2005). The charge transfer mechanism that allows the electron to be promoted from the metal into the diamondoid LUMO is very efficient, and electrons are emitted almost instantaneously with an upper bound of a few femtoseconds. This suggests that if suitable pulsed light sources are available, diamondoid SAMs could be used as ultrashort (~fs), high-brilliance pulsed electron sources for use in electron microscopy, electron beam lithography, field-emission flat-panel displays, and photocathodes.

Investigations of diamondoids up to [121]-tetramantane also showed that the dielectric constant κ is about half the one of bulk diamond ($\kappa \approx 5.6$), ranging from 2.46 for adamantane to 2.68 for tetramantane (Clay et al., 2008). This suggests that diamondoids could be new candidates as low-κ materials for microelectronics applications.

Using the negative electron affinity properties of diamondoids, a photocathode device using diamondoid and bromide films has been demonstrated recently (Clay et al., 2012). The photoyield of the diamondoid/bromide films was higher and more stable compared to the diamondoid-only films.

It has also been shown that the bandgap of diamondoids can be tuned by either changing their size or changing functional groups (Fokin and Schreiner, 2009). The authors conducted an extensive density functional theory (DFT) study for elucidating the effect of different doping strategies on the bandgap of diamondoids. The approaches they investigated were the enlargement and changing of the shapes/morphologies of the particles, the effect of C–H bond substitutions with various functional groups (external doping), and finally the effect of incorporating heteroatom functionalities of one or more alkylated terminations such as CH or CH_2 fragments (interstitial or internal doping).

They found that by increasing the size of diamondoids from $C_{10}H_{16}$ (adamantane) to about 2 nm ($C_{286}H_{144}$), the bandgap decreases from 6.7 eV, which is still larger than the bandgap of bulk diamond (5.5 eV). It was also found that it is the size and not the morphology that influences the bandgap most.

In their paper, the authors discovered that bandgap tuning through external (by C–H bond substitution) or internal doping (by replacing CH or CH_2 groups) is nonadditive for the same dopant. This means that, for instance, doubling the number of same functional groups does not lead to a linear decrease in the bandgap. Functionalization by attaching electron-donating and electron-withdrawing groups (push-pull doping) was found to be most effective and permitted to reduce the bandgap of diamondoids to that of bulk diamond.

Sasagawa et al. (Sasagawa and Shen, 2008) studied the effect of clustering on the bandgap. DFT calculations showed that in contrast to bulk diamond, crystalline adamantane has a direct bandgap. As the transition of electrons across the bandgap can accompany the absorption and emission of photons, the efficiency of this process can be expected to be far better in crystalline diamondoid materials compared to indirect-bandgap materials.

4.5 Possible Future Applications of Diamondoids

It is expected that because of their well-defined structure and high stability, diamondoids could also be used in many different other applications in the future.

In the following section, we will present a few examples of promising research involving the use of diamonds or nanodiamonds that could potentially be replaced by diamondoids.

This research includes photonics but also the field of quantum communication and computing. Macroscopic diamond has been hailed as a material that can match the requirements of the second quantum revolution (Aharonovich et al., 2011). As will be shown in detail in the following section, sufficiently large diamondoids that contain nitrogen vacancy (NV) centers could replace currently used diamonds for some of these applications. Figure 4.11 illustrates a diamond unit cell containing such an NV center.

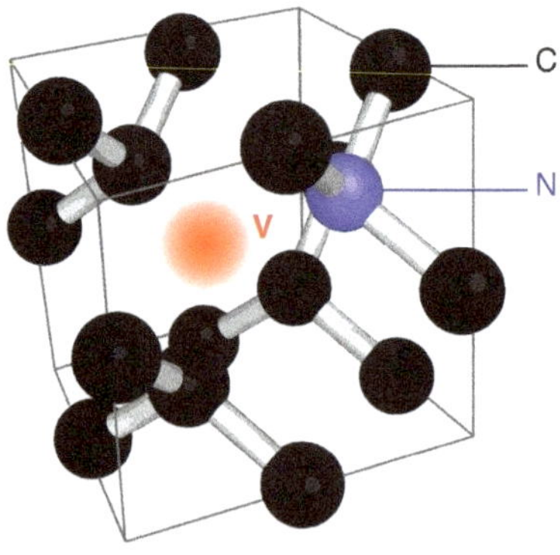

Figure 4.11 Crystallographic structure of nitrogen vacancy center in diamond. The vacancy (V) is surrounded by a single nitrogen (N) and three carbon atoms (C). Adapted from Mizuochi et al. (2012).

In subsequent sections, we will also briefly discuss the possible use of diamondoids as molecular building blocks in exotic sounding applications such as nanorobots and molecular machines.

While actual implementation of nanorobots and molecular machines still has to be demonstrated, we believe that the structural stability and well-defined shapes of diamondoids make them ideal building blocks for such applications.

4.5.1 Biotechnology

One of the key breakthroughs in understanding cell processes has been the real-time imaging of cell dynamics. Recent studies have demonstrated 3D tracking of single fluorescent nanodiamonds inside cells, as well as preliminary experiments of imaging living cells, drug delivery, and targeted bioimaging. As mentioned in the previous section, the recent discovery of new ultrabright colour centres with narrow emission in the near-infrared region is significant for such tracking techniques, as these wavelengths do not interfere with cell autofluorescence. However, no mass production technique yet exists that can incorporate such bright emitters into nanodiamonds (Aharonovich et al., 2011).

For example, chemically modified diamondoids have been suggested as potential biosensors for DNA testing (Sivaraman and Fyta, 2014). So far, mainly nanodiamonds have been used for investigating cellular mechanisms, which is evident from the large number of texts in the literature.

4.5.2 Quantum Computing and Communication

In recent years, advances for future access of both quantum communication and computation have increased considerably.

Conventionally, to entangle spins over large distances, it has been necessary to work at very low temperatures of a few Kelvin. However, in recent years, alternative approaches have been proposed. It has been found that diamonds containing permanent spins, caused by NV centers, that is, the presence of nitrogen impurities and carbon vacancies, permit spin entanglement at room temperature. A schematic of a diamond crystal lattice containing such an NV center is shown in Fig. 4.11. This NV defect represents a permanent spin that has been shown to be useful for different applications, such as small magnetic probes.

Up to now, the potential for application in quantum communication and quantum computing has been shown for macrodiamonds with sizes of a few millimeters. However, nanodiamonds possessing a single NV have also been reported (Aharonovich et al., 2011). One could imagine that such diamonds that are currently of several millimeters could be replaced by nanodiamonds or even higher-order diamondoids. However, currently, it is also not known how large such diamondoids would have to be so that their structure and consequently also the NV color centers are sufficiently stable for spin entanglement. Currently there are no studies that have investigated the effect of diamondoid size on the stability of NV color centers, that is, to determine what would be the smallest necessary size of a diamondoid to accommodate an NV center. It would be interesting, both from a theoretical and from a practical point of view, to investigate this in the future.

In addition to quantum computing and communication applications, diamondoid-like porous networks could also be used for molecular shape sorting (Mitra et al., 2013).

4.5.3 Magnetometry

Investigating magnetism at the nano- and atomic scale is a key issue both for understanding fundamental physical properties of matter and as the enabling ingredient for magnetism-based data storage and spintronic devices (Degen, 2008).

The NV centers in diamond have been shown to be useful for measuring very small magnetic fields.

A disadvantage of magnetic resonance force microscopy (MRFM) is that it only performs well at cryogenic temperatures and in vacuum, which is not compatible with many biological (and other) applications. Therefore we are trying to come up with other ideas that might allow us to do the same experiments under ambient conditions.

In scanning diamond magnetometry, quantum optics is used to detect minute magnetic signals from samples via the fluorescent light emitted by NV centers in diamond. As already mentioned, such NV centers carry a spin, which is very sensitive to magnetic fields. As the optical rate of emission depends on the defects spin state, small magnetic fields can be measured by looking at the fluorescence intensity (Degen, 2008; Taylor et al., 2008).

Placing a diamond defect in a nanocrystal and attaching it to a scanning probe allows performing imaging at the nanometer scale. The big advantage is that such scanning diamond magnetometers have the capability to measure the dipole field of single-electron spins, or monolayers of organic materials containing proton spins, at room temperature. The combination of diamond sensors with magnetic resonance spectroscopy allows realizing of the chemical analysis of surfaces (organic, metallic, etc.) with nanometer resolution.

4.5.4 Nanorobots and Molecular Machines

Because of their high thermal stability compared to other organic molecules such as alkanes, diamondoids might be useful as building blocks for nanorobots and in molecular machines.

Rotary dynamics at the molecular level hold promise for various applications, and the effects of rotational symmetry on the dynamics of a molecular rotor consisting of phenylene and diamantane has been investigated (Karlen et al., 2005).

The advantage of diamondoids is that compared to other rigid molecular structures that could be candidates for molecular rotors, they possess higher cylindrical symmetries. For example, in the case of structures possessing Cn symmetries with $n > 2$, there are n energy minima and $360^\circ/n$ angular displacements. An example of such a molecular rotor is displayed in Fig. 4.12.

Based on different diamondoids, basic toolsets for diamond mechanosynthesis were defined. These include parts for abstracting hydrogen and also to attach other molecules. A minimal set of such a diamondoid-based mechanosynthesis toolset is presented in Fig. 4.13, and Fig. 4.14 shows the application of the HABst tool for removing a hydrogen atom from a (111) diamond surface.

In the wake of nanotechnology, bottom-up approaches have also been proposed to be applicable for medical applications,

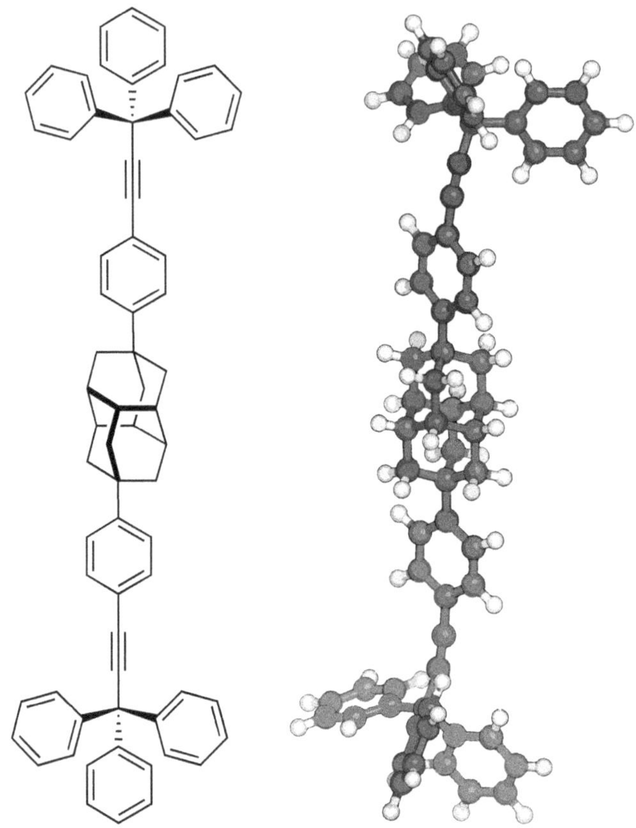

Figure 4.12 Diamondoids as building blocks for molecular robots. Molecular structure of 4,9-bis(4-iodo-phenyl) diamantane that can serve as a rigid rotor.

so-called nanomedicine. One of the hallmarks of nanomedicine, as proposed by several researchers, is the treatment of diseases directly in the body or cells. To make this possible, the use of nanorobots has been proposed (Patel et al., 2006). Because of their advantageous properties and as an extension of diamondoid-based mechanosynthesis, it has also been suggested that diamondoids could be used to fabricate nanorobots that could be used for medical applications (Freitas, 2009).

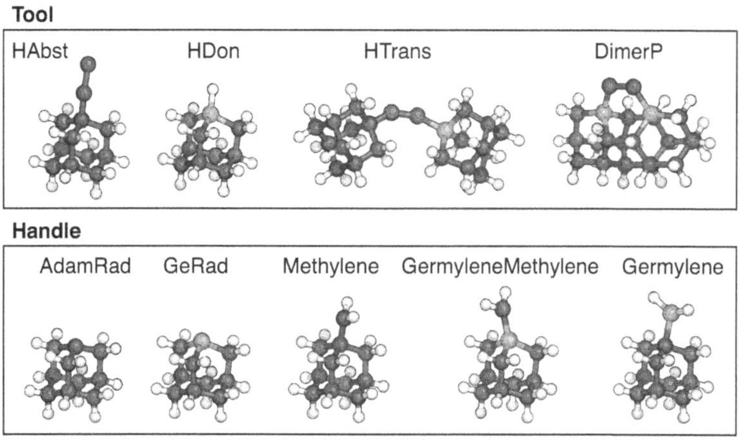

Figure 4.13 Minimal diamondoid toolset. Adapted from Freitas (2009).

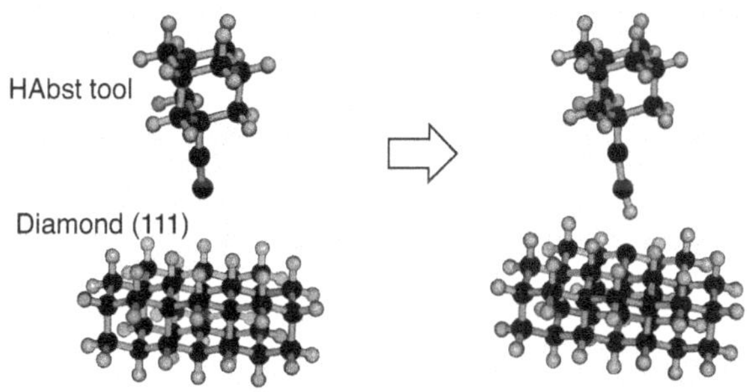

Figure 4.14 Hydrogen abstraction via mechanosynthesis. A hydrogen termination is removed from a diamond (111) surface by the use of the HAbst tool. Adapted from Freitas (2009).

4.6 Summary

As has been outlined in this chapter, the well-defined structure of diamondoids, which distinguishes them from nanodiamonds, makes them predestined for use as molecular building blocks in a wide range of applications. The highest potentials for the application of diamondoids is certainly in medicine and biotechnology, where adamantane derivatives have already been used to treat Alzheimer's and Parkinson's diseases and have also acted as the bases for other drugs.

In addition to these applications, diamondoids that contain structural defects, namely NV centers, could be useful for spintronic applications and quantum computing and quantum communication. However, while certain physical properties of diamondoids have already been demonstrated experimentally, the main problem is their (non)availability, especially that of higher diamondoids with $n \geq 4$, in large quantities.

In the following part, the methods used up to now for obtaining diamondoids, either by isolation from natural gas and oil reservoirs or by organic synthesis, are presented.

Part II

Isolation and Organic Chemical Synthesis of Diamondoids

Chapter 5

Occurrence and Isolation of Diamondoids from Natural Gas and Oil Reservoirs

The topic of this chapter is the isolation of diamondoids from natural sources, namely gas and oil reservoirs. We briefly present a short history of their discovery from the 1930s to the present date and the techniques necessary for their isolation and purification that were developed following the isolation of higher diamondoids up to undecamantane in the 1990s.

5.1 Occurrence of Diamondoids in Natural Gas and Oil Reservoirs

The first member of diamondoids, adamantane, was first found in crude oil in 1933 in Czechoslovakia in the region of Hodonin. Since then, diamondoids have been discovered in a variety of gas and oil reservoirs, including Japan (Shimoyama and Yabuta, 2002). While diamondoids are thermally stable, it has been found that over time, and if the temperature is above the peak diamondoid formation

Diamondoids: Synthesis, Properties, and Applications
Sven Strauss and Kazuo Terashima

ISBN 978-981-4745-18-5 (Hardcover), 978-1-315-36466-7 (eBook)
www.panstanford.com

temperature of about 340 K, they can decompose again (Schoell and Carlson, 1999).

To understand the formation of diamondoids in natural sources, attempts have also been made to draw parallels with nanodiamond formation using explosives (Pichot et al., 2013). Since diamondoids often lead to fouling of the piping during oil or gas exploration, that is, when they form solid condensates, over the years methods to isolate and separate them from crude oil have been developed (Huang et al., 2011).

The importance of diamondoids in natural fossil sources, including oil, has long been recognized, and also the implication of the presence of diamondoids on the structure of coal has been investigated (Aczel et al., 1979).

There have also been attempts to understand the reactions of diamondoids, especially alkylated adamantanes, in sediments, and it has been suggested that carbon surface reactions play an important role (Berwick et al., 2011).

Besides natural oil and gas reservoirs, traces of diamondoids have also been found in meteorites, which would imply that diamondoids were also formed in the early solar system (Dai et al., 2002). Moreover, signs of diamondoids have also been detected in carbon-rich proto-planetary nebulae (Hill et al., 1998). Finally, it has been suggested that diamonds, including diamondoids, could be of presolar origin, which was found by comparing samples of meteoritic and terrestrial origin by high-resolution transmission electron microscopy (Daulton et al., 1996).

Catalytic and thermal cracking of pure hydrocarbons in oil and gas reservoirs has been investigated since the 1940s (Greensfelder et al., 1949). After the first discoveries of adamantane in oil sources, diamantane also could be isolated from petroleum (Hala et al., 1966).

Finally, it has also been surmised that not only diamondoids but also macrodiamonds could be found in crude oil (Hopf, 2003).

In normal gas and oil reservoirs, the contents of diamondoids is of the order of ppm or even sub-ppm (Liang et al., 2012), but the concentration increases with the degree of oil cracking, which is associated with depths within the earth of about 5 km and temperatures of 150°C–175°C, the so-called oil deadline. Because of their high thermal stability, diamondoids can serve as indicators for the amount of oil cracking. Dahl et al. (Dahl et al., 1999) investigated the concentration increase of methyldiamantane as a function of cracking of oil to gas and pyrobitumen increases diamondoid concentration. The concentration of methyldiamantanes in uncracked oils—the diamondoid baseline—is on the order of 1–10 ppm, depending on the source(Dahl et al., 1999).

Biomarkers are molecular fossils that possess recognizable carbon structures of natural products and that are synthesized by oil precursor organisms such as bacteria and algae.

5.2 Formation of Diamondoids in Natural Sources

In this section, we will give a brief overview of the processes that lead to the formation of diamondoids in natural gas and oil reservoirs.

As will be explained in more detail in later sections, with the exception of the lower diamondoids adamantane and diamantane, currently the main sourced of higher diamondoids are natural oil and gas reservoirs, while especially conventional organic chemical synthesis methods have been demonstrated to be limited by the many different possible reaction pathways, reaction intermediates, and local minima on the potential energy surface.

Understanding the natural processes of diamondoids in natural sources and the conditions under which the formation of diamondoids is highest might also help to improve artificial synthesis of diamondoids.

Both gas and oil are produced from organic matter. Almost all, that is, more than 99% of the organic matter which is produced on land or in the oceans is dissociated by direct oxidation or by microbiological processes.

Figure 5.1 illustrates schematically the carbon cycle and the region where the main precursor of diamondoids, kerogens, are formed.

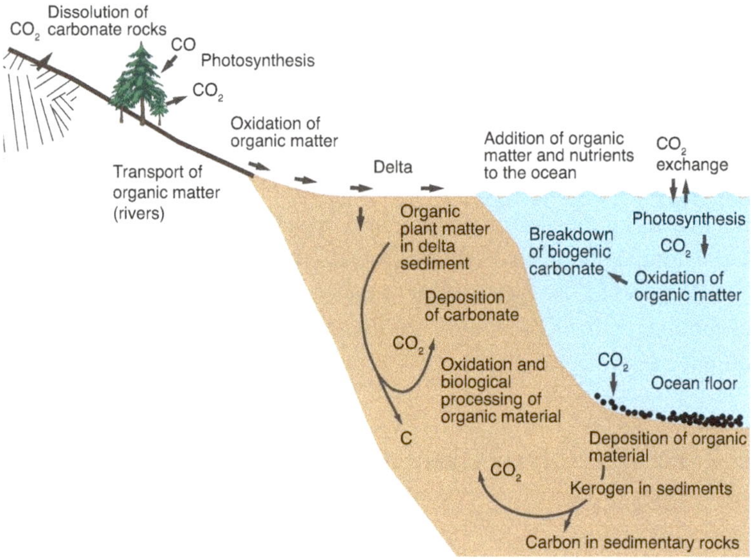

Figure 5.1 Illustration of the carbon cycle. Carbon from organic matter and carbonate rocks are the major sinks for carbon (CO_2). The rate of precipitation of carbonate in the ocean by organisms is limited by the supply of Ca^{++} and Mg^{++} from weathering of silicate rocks brought in by rivers. The illustration also shows where the main precursors of diamondoids, kerogens, are formed. After Bjørlykke (2010).

As organic material, both from plants and from organisms, becomes buried by the accumulation of sediments, water is gradually expelled during compaction. Complex organic compounds like proteins are broken down into amino acids, while carbohydrates transform into sugar compounds. These smaller compounds are able to recombine to form larger compounds, for example, by amino acids reacting with carbohydrates (melanoid reaction) (Bjørlykke, 2010). As this type of polymerization proceeds, the proportion of simpler soluble organic compounds diminishes at depths of a few tens of meters down in the sediment. These newly formed complex organic structures are called kerogens. In short, kerogen is a collective term for organic material that is insoluble in organic solvents, water, or oxidizing acids. Kerogen consists of very large molecules and is therefore similar to macromolecules or polymers. When kerogens

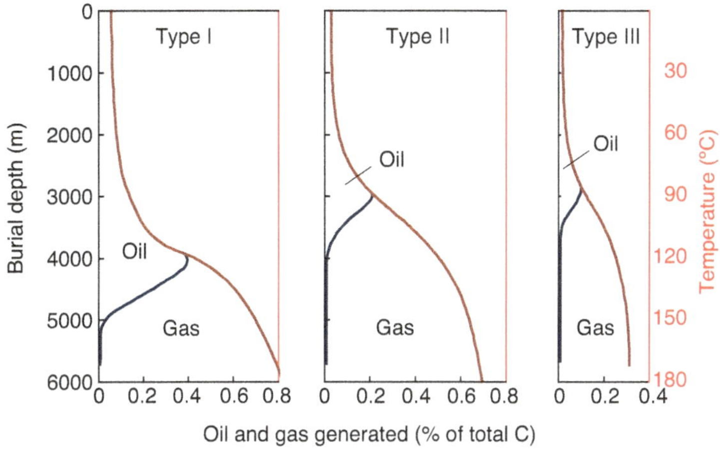

Figure 5.2 Oil and gas formation as a function of temperature and depth. Data adapted from Vandenbroucke (2003).

are exposed to elevated temperatures, the large molecules crack into smaller ones, predominantly petroleum. At temperatures of about 100°C, long geological times are required for this process. In reservoirs that are sinking rapidly, the exposure time is shorter and oil generation can only start at about 140°C to 150°C. Figure 5.2 indicates the depth and temperature conditions for the formation of oil and gas.

Figure 5.3 displays the variation of the diamondoid content (adamantane, diamantane, and derivatives) as a function of EasyRo in percent. EasyRo (also written as EASY%R_o) is a model (Sweeney and Burnham, 1990) that allows predicting the level of vitrinite[a] maturation, that is the change in the level of vitrinite in a given sedimentary rock as a function of time and temperature. As can be seen in the figure, the maximum is reached at about 2.0 EasyRo.

It has also been found that diamondoids are more abundant in reservoirs that contain clay minerals, especially montmorillonite (K10), whose structure is depicted in Fig. 5.4. To investigate the influence of the presence of clay minerals on the formation of

[a] Vitrinite is an organic coal and kerogen component and has a shiny appearance similar to that of glass.

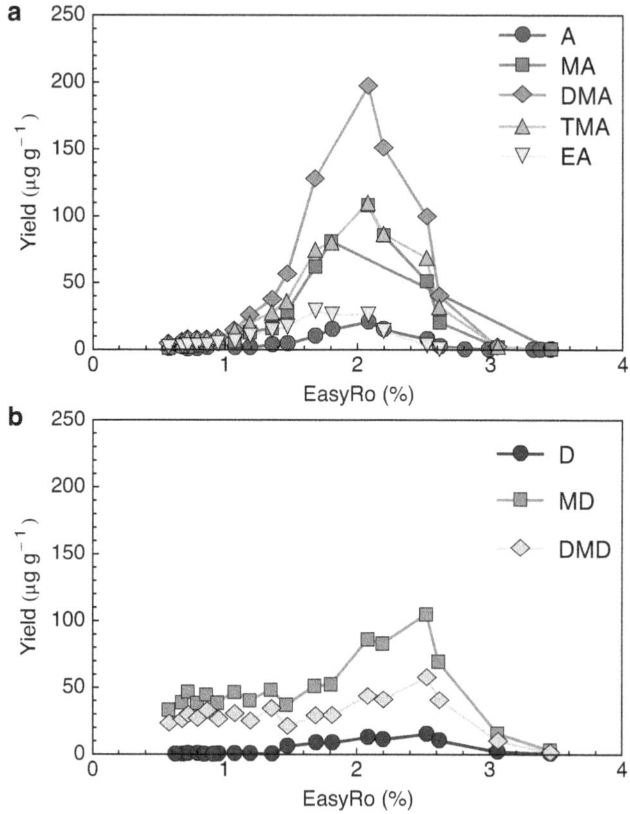

Figure 5.3 Variation in the yields of diamondoids with EasyRo (%) in the oil- cracking experiment. The yields are given in $\mu g\,g^{-1}$ oil. (a) Adamantanes and (b) diamantanes. A, adamantane; MA, methyladamantanes; EA, ethyladamantanes; DMA, dimethyladamantanes; TMA, trimethyladamantanes, D, diamantane; MD, methyldiamantanes; DMD, dimethyldiamantanes. Data adapted from Fang et al. (2012).

diamondoids, Moldowan and coworkers conducted a series of experiments. These consisted of using diamondoid precursors such as kerogens and sealing them in vessels that were heated to temperatures typical of oil formation.

However, while diamondoids can grow at temperatures between 150°C and 250°C, if they are outside this temperature window, then

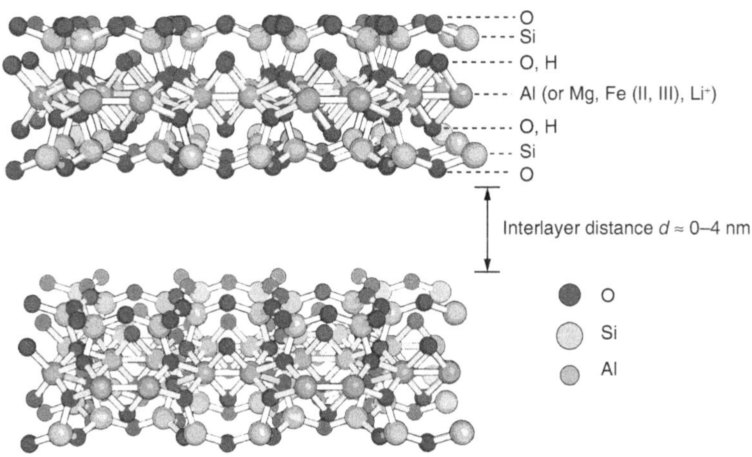

Figure 5.4 Crystal structure of montmorillonite, K10. Depending on the type of montmorillonite, the layers can exhibit different chemical compositions. The interlayer distance *d* depends on the type and amount of absorbed molecules.

they can decompose again. In addition, it has also been found that diamondoids can be decomposed by certain bacteria.

5.3 Isolation of Diamondoids from Gas and Oil

From a practical point of view, for oil and gas exploration, the presence of diamondoids is not desirable, as they often form condensates, clogging piping systems. In order to separate diamondoids from oil or gas condensates, several steps are necessary (Dahl et al., 2003a). First, the oil condensates are distilled under vaccum at temperatures above 345°C. In a next step, pyrolysis at $T = 400°C–450°C$ is used to remove nondiamonds from the mixture. Then, silica gel liquid chromatography is employed to remove aromatic and polar compounds, and finally, reverse-phase high-performance liquid chromatography (HPLC) is used to separate the diamondoids, and shape selective HPLC can be used to differentiate diamondoid enantiomers. The flowchart illustrating the diamondoid separation process is illustrated in Fig. 5.5.

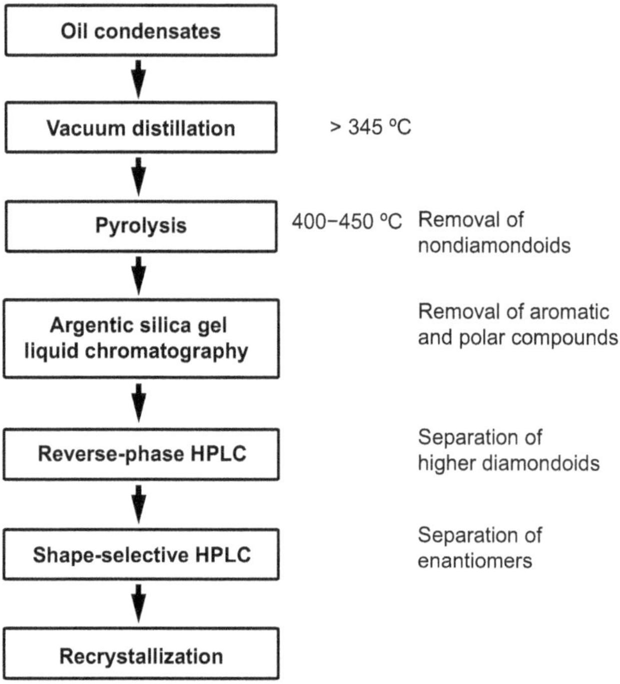

Figure 5.5 Schematic of the separation process for obtaining single crystals of higher diamondoids. Oil condensates are concentrated by vacuum distillation at temperatures above 345°C. Volatile components and nondiamondoids are cracked by pyrolysis conducted in the temperature range of 400°C–450°C. In the next step, polar and aromatic compounds are removed by silica-gel-based column chromatography. Individual types of higher diamondoids are isolated by reverse-phase HPLC, enantiomeric diamondoids being separated by shape-selective HPLC. The separated fractions are finally recrystallized.

5.3.1 Alternative Purification Methods

As an addition to the original isolation and purification process (Dahl et al., 2003a), a more advanced refinement method using a two-zone furnace for the growth of large single crystals of diamondoids by the so-called vapor transport technique has been developed (Iwasa et al., 2010). In this method, by taking advantage of the low binding energy due to weak van der Waals forces and the low sublimation temperatures of diamondoids, the sample is

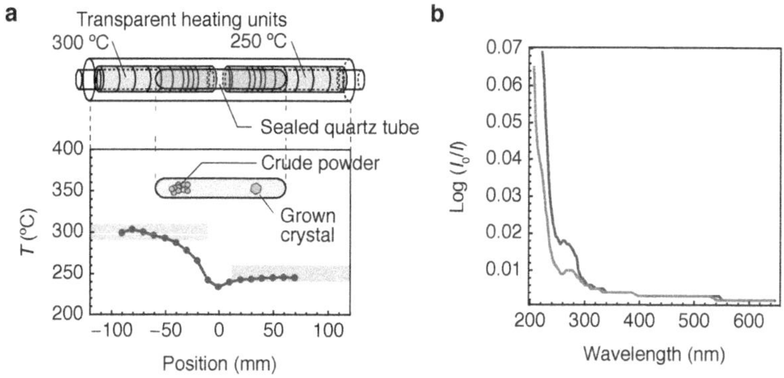

Figure 5.6 Schematic of the two-zone furnace used for refinement and crystallization of diamondoids. (a) Two-zone furnace consisting of a quartz tube and a thin wire that acts as a heating element. (b) Absorbance spectra of commercial and refined diamantane dissolved in *n*-hexane. The reduced intensity of the absorbance peak at $\lambda \approx 260$–280 nm in the refined sample indicates the decrease of impurities. Schematic of furnace and data adapted from Iwasa et al. (2010).

first brought in the vapor phase by heating of one end of the tube. The particles brought into the vapor phase then deposit in different zones along the tube.

This method allows to obtain high-quality diamondoid crystals without having to resort to other purification techniques such as multistage distillation or high-performance liquid chromatography (HPLC), and it is possible to obtain diamondoid single crystals with volumes up to $\sim 1\,\text{cm}^3$ and to reduce the number of impurities considerably.

Figure 5.6a shows a schematic of the two-zone furnace and the temperature distribution along the axis of the sealed quartz tube. The crystals recrystallize in the zone with the lowest temperature. Figure 5.6b shows a comparison between absorption measurements of a commercially available and a refined diamantane sample.

Chapter 6

Approaches for the Organic Synthesis of Diamondoids

In this chapter, we give a short overview of the currently known routes for synthesizing diamondoids using chemical synthesis approaches, without the pretension to be fully exhaustive, as there is a large body of research literature available, to which the reader will be pointed. As will be explained in detail, the complexity in synthesizing diamondoids grows exponentially with an increasing cage number, and to date, there are only methods that allow obtaining all the lower diamondoids—adamantane, diamantane, and triamantane—and one higher diamondoid, anti-tetramantane. We also summarize the currently known main procedures for their functionalization, before closing this chapter with a short section on the difficulties and limits related to the synthesis of diamondoids by conventional organic chemical synthesis.

Diamondoids: Synthesis, Properties, and Applications
Sven Strauss and Kazuo Terashima

ISBN 978-981-4745-18-5 (Hardcover), 978-1-315-36466-7 (eBook)
www.panstanford.com

6.1 A Brief History of the Isolation and Organic Synthesis of Diamondoids

As mentioned in Chapter 2, the structure of adamantane based on the diamond lattice was predicted already at the end of the 19th century.

An extensive account on the history of the synthesis of lower diamondoids is given in the paper by Schwertfeger et al. (2008), and here, we briefly summarize the main points. As has already been mentioned in Chapter 5, the structure of adamantane was already predicted before it was isolated for the first time from crude oil in Hodonin in former Czechoslovakia in 1933.

The first artificial synthesis of adamantane was reported by (Prelog and Seiwerth, 1941a), but the yield was very low, only on the order of a few percent. Especially in the 1950s and 1960s, research on the synthesis of diamondoids became very active, and several methods were developed to obtain diamondoids up to antitetramantane.

Following the synthesis of the lower diamondoids in reasonable yields, the organic synthesis of higher diamondoids was the topic of further intensive efforts up to the early 1980s. However, it could be demonstrated that while lower diamondoids can be synthesized by carbocation equilibration reactions, this approach does not work for higher diamondoids.

Other problems that prevented successful synthesis of higher diamondoids were the lack of suitable precursors, the rapidly growing number of possible isomers, intermediates being trapped in local energy minima, and the formation of unwanted side products. As a result, research on carbocation-mediated syntheses of diamondoids beyond tetramantane was abandoned at the beginning of the 1980s (Osawa et al., 1980), and up to now, only one of the higher diamondoids, [121]tetramantane, has been synthesized, but the reaction is complex and the reaction yields very low (Burns et al., 1978).

The interest of diamondoids was only renewed when it was reported that diamondoids with a number of cages (n) of up to 11 were isolated from crude oil (Dahl et al., 2003a) (see

Chapter 5). Since then, work on diamondoids, both experimental and theoretical, has increased again.

In the next section, we will discuss the approaches used for the organic synthesis of diamondoids in detail.

6.2 Conventional Organic Chemical Synthesis of Diamondoids

As has been outlined in the previous section, since the discovery of the first diamondoid members adamantane and diamantane, many attempts have been undertaken to obtain diamondoids synthetically.

In this section, we treat the synthesis of diamondoids from adamantane up to tetramantane in more detail, but we also refer the interested reader to detailed works in the literature. As will be seen, the complexity of the diamondoid reactions increases with the number of diamondoid units, while the reaction yields, especially for diamondoids above diamantane, decrease.

What will also become evident from this section is that both adequate precursors and catalysts are crucial for obtaining the desired structures. We close this section with a brief explanation on why, currently, there are no known successful approaches that would allow the facile synthesis of diamondoids larger than tetramantan in sufficient yields.

6.2.1 Synthesis of Adamantane

The structure of adamantane was already predicted at the end of the 19th century, but it was only in 1933 when it was first isolated from oil in Hodonin in today's Czech Republic. Since the discovery, various attempts have been made to synthesize adamantane by different means and while the first approaches gave only small yields, more effective reaction schemes were developed over time. In this section, we will present the currently most commonly used approaches for the synthesis of adamantane.

While originally, all the smaller diamondoids were isolated from crude oil, nowadays adamantane is predominantly synthesized by

Figure 6.1 Reaction scheme for the synthesis of adamantane. Cyclopentadiene serves as a precursor, which is then hydrogenated to tetrahydrodicyclopentadiene using platinum dioxide (PtO_2) and diethyl ether (Et_2O). Using a strong Lewis acid ($AlCl_3$), tetrahydrodicyclopentadiene can then be rearranged to form diamantane.

conventional organic synthesis, and its production is on the order of several tons per year.

The first report on the synthetic fabrication of adamantane dates from 1941 (Prelog and Seiwerth, 1941b). In the same year, the same authors further refined their approach, which allowed them to increase the reaction yield from 2% to higher values (Prelog and Seiwerth, 1941a).

There were also a few other attempts to synthesize adamantane more effectively, but these approaches always required several reaction steps and the yields were always only on the order of a few percent. Then, an approach based on so-called stabilomeric rearrangement of tetrahydrodicyclopentadiene by Lewis acids allowed the synthesis of adamantane (Schleyer, 1957). This method involves two reaction steps, which are illustrated in Fig. 6.1. This method allowed reaching reaction yields of 12–13%.

6.2.2 Synthesis of Diamantane

The diamantane structure was chosen as the official emblem of the XIXth International Congress of Pure and Applied Chemistry, which was held in London in 1963, and the synthesis of congressane', as it was called, was sugggested as a challenge for the participants of the congress. Following this challenge, a first approach was first demonstrated two years after the congress (Cupas et al., 1965). Figure 6.2 illustrates two possible synthesis methods for the synthesis of diamantane. In both approaches, norbornene is first converted photochemically in the presence of photosensitizing agents (symbolized by $h\nu$ in Fig. 6.2) to a norbornene dimer

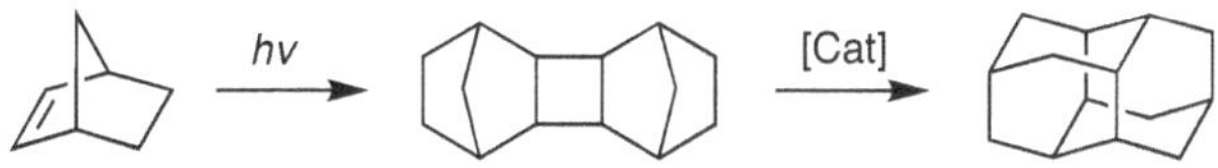

Cat:
1) $AlCl_3$: 1%
2) $AlBr_3$ / *t*-BuBr: 10–11%

Figure 6.2 Schematic of diamantane synthesis. The starting material is norbornene, which is converted photochemically first to the norbornene dimer (intermediate product between norbornene and diamantane). When treated with a catalyst ("Cat"), in this case $AlCl_3$ (method 1), tar but also diamantane is produced. The yield of this reaction is on the order of 1%. When instead using $AlBr_3$ and *tert*-butyl bromide (*t*-BuBr, method 2), reaction yields up to 11% are obtained.

($C_{14}H_{20}$), whose structure is shown in the middle of Fig. 6.2. When treated with $AlCl_3$ (method 1), some isomers of this norbornene dimer can give large quantities of tar, but others also give small amounts of diamantane, of the order of 1% (Cupas et al., 1965). By replacing $AlCl_3$ with a mixture of $AlBr_3$ and *tert*-butyl bromide (*t*-BuBr, method 2), the reaction yields can be increased up to 11% (Gund et al., 1970; Williams et al., 1966).

In addition to the reactions described in Fig. 6.2, various alternative synthesis routes were explored in order to achieve even higher reaction yields of diamantane. Figure 6.3 gives a few more examples of such synthesis approaches, which are based on basically three different precursors. The first is based on using cyclopentadiene and the formation of a $C_{14}H_{20}$ isomer as reaction intermediate using different types of approaches (labeled methods 1–4 in Fig. 6.3a) (Gund et al., 1970). In Fig. 6.3b, the synthesis of diamantane is achieved by using the so-called "Katz dimer" as a precursor (Mrowca and Katz, 1966). Finally, Fig. 6.3c lists different reaction schemes where "Binor-S" was used as a starting material.

Finally, Fig. 6.4 illustrates a few more alternative approaches for the diamondoid synthesis. Figure 6.4a shows two possible methods based on the use of cycloheptatriene as a precursor, which is then dimerized with the aid of titanium complex catalyst. The following step involves hydrogenation in the presence of platinum as a catalyst.

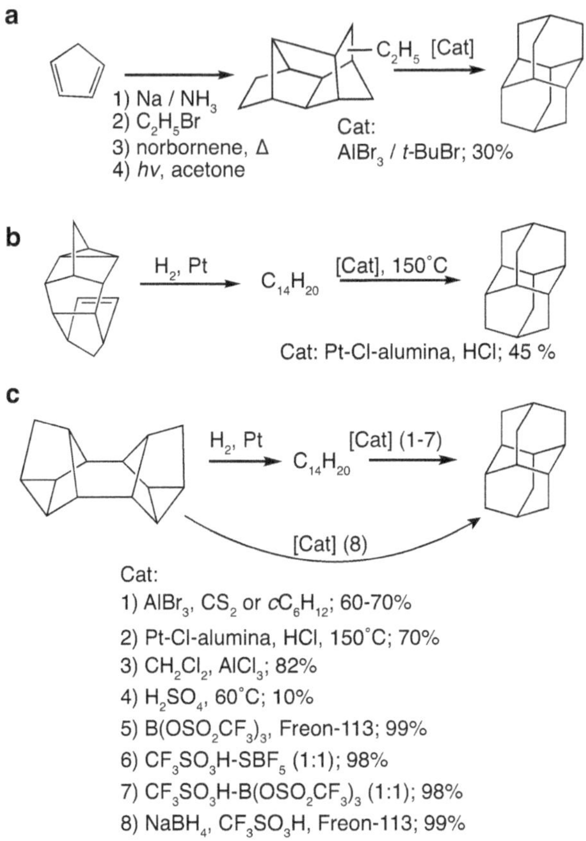

Figure 6.3 Alternative approaches for the synthesis of diamantane using different types of precursors and catalysts. (a) Cyclopentadiene as a precursor. (b) Katz dimer as starting material. (c) Binor-S as precursor and using different types of catalyists (methods 1–8). The percentages listed with the catalysts indicate the respective reaction yields and the symbol Δ indicates the use of heat. Reaction schemes adapted from Schwertfeger et al. (2008).

This approach leads to reaction yields of up to 89% of diamantane (Schwertfeger et al., 2008).

The reaction scheme in Fig. 6.4b is based on the use of spiro[2.4]hepta-4,6-diene as a precursor (Schwertfeger et al., 2008). However, this approach does give diamantane yields of only 13% or

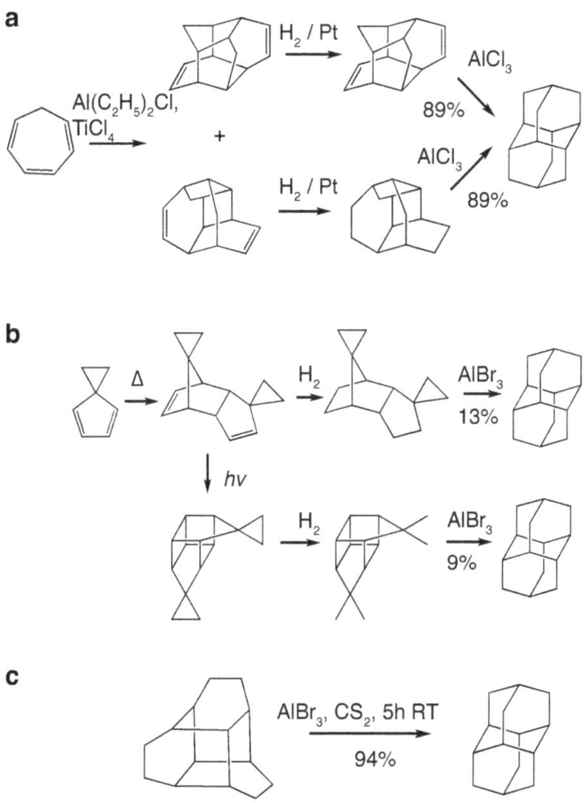

Figure 6.4 Alternative synthesis approaches for the synthesis of diamantane. (a) With cycloheptatriene as precursor. (b) Spiro[2.4]hepta-4,6-diene as starting material. (c) Using a pentacyclotetradecane (D_3-tritwistane) as precursor. The percentages indicate the reaction yields. Reaction schemes adapted from Schwertfeger et al. (2008).

9%. The reaction scheme depicted in Fig. 6.4c is based on the use of a pentacyclotetradecane D_3-tritwistane, which allowed achieving reaction yields of up to 94% (Nakazaki et al., 1981).

In addition to the diamantane synthesis based on more or less easily available precursors, there have also been attempts to synthesize diamantane in a stepwise fashion using the next smaller diamondoid—adamantane—as a precursor. Using this approach, diamantane could be synthesized from adamantane in several steps that involved the functionalization of the adamantane structure and subsequent ring closures and rearrangements (Farcasiu et al., 1977). In principle, this approach could be extended to achieve the synthesis of higher diamondoids, but because of the many reaction steps involved, it has turned out to be not very practical.

6.2.3 Synthesis of Triamantane

As for the synthesis of adamantane and diamantane, for triamantane, over the years many different reaction mechanisms have been investigated. The most common methods that give the highest reaction yields are summarized in Figs. 6.5 and 6.6. Also for triamantane, the reaction yields highly depend on the nature of the available precursors and catalysts.

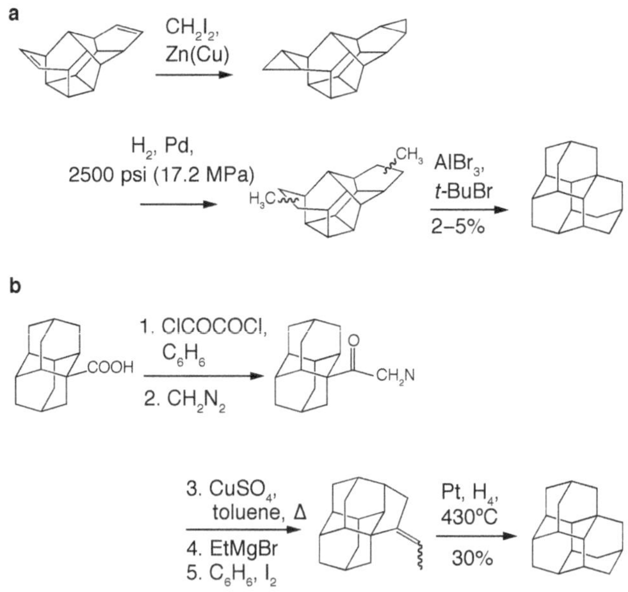

Figure 6.5 Possible reactions for the synthesis of triamantane. (a) Use of a cyclooctatetraene dimer as precursor. (b) With 1-diamantanecarboxylic acid as a starting material. The percentage values indicate the reaction yields and the symbol Δ indicates the use of heat. Adapted from Schwertfeger et al. (2008).

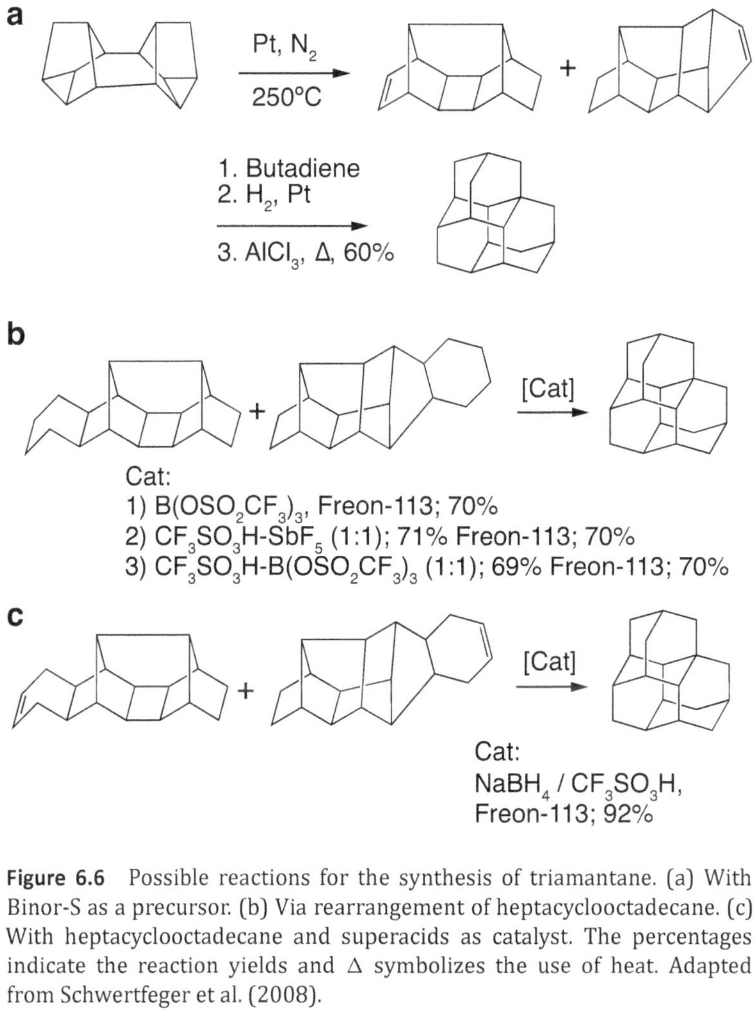

Figure 6.6 Possible reactions for the synthesis of triamantane. (a) With Binor-S as a precursor. (b) Via rearrangement of heptacyclooctadecane. (c) With heptacyclooctadecane and superacids as catalyst. The percentages indicate the reaction yields and Δ symbolizes the use of heat. Adapted from Schwertfeger et al. (2008).

The most common approach consists in the elaboration of an isomer of triamantane ($C_{18}H_{24}$), whose structure consists of seven rings and which is then rearranged to yield the triamantane structure. One example for the synthesis of triamantane consisted in the use of a cyclooctatetraene dimer (Fig. 6.5a). An alternative approach consists in the use of 1-diamantanecarboxylic acid as a starting material, whose reaction scheme is depicted in Fig. 6.5b (Burns et al., 1975). This method gave a reaction yield up to 30%.

Another synthetic route for the fabrication of triamantane consists in the use of S-Binor as a precursor, as depicted in Fig. 6.6a, which gave yields of up to 60% (Hamilton et al., 1976; Hollowood et al., 1980). It has also been demonstrated that triamantane can be obtained via catalysis of the rearrangement of heptacyclooctadecanes in the presence of superacids (Farooq et al., 1988). This approach is illustrated in the reaction scheme of Fig. 6.6b. To achieve the synthesis, different types of catalysts were used (labeled methods 1–3), which resulted in reaction yields of up to 70%. Finally, it was also shown that triamantane can be synthesized from heptacyclooctadecanes using $NaBH_4/CF_3SO_3H$ superacids as catalyst in a Freon-113 solution (Olah et al., 1989) (cf. Fig. 6.6c).

6.2.4 Synthesis of Tetramantane

In addition to the lower diamondoids adamantane, diamantane, and triamantane, extensive attempts were done to synthesize higher diamondoids, starting with tetramantane. However, although many groups were active in this field, up to now, only one structural isomer of tetramantane, [121]tetramantane (also called antitetramantane), has been synthesized. Figure 6.7 shows the synthesis steps for obtaining antitetramantane (Burns et al., 1978), which is based on the use of 1,6-diamantanedicarboxylic acid as a precursor. As for triamantane, the synthesis of tetramantane has also been attempted by using a solid catalyst-mediated rearrangement, namely on platinum (Burns et al., 1975, 1976), which gave reaction yields of up to 10%.

After the unsuccessful attempts to extend the approaches of lower diamondoid synthesis, research on diamondoids waned in the 1980s. Renewed interest in diamondoids only started after the discovery of higher diamondoids up to undecamantane in natural oil and gas reservoirs (Dahl et al., 2003a) at the beginning of the 2000s.

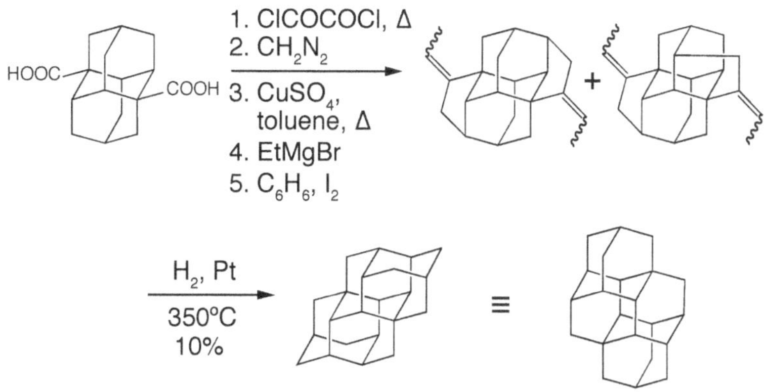

Figure 6.7 Schematic of the reaction pathway of [121]tetramantane (anti-tetramantane). The synthesis is based on the use of 1,6-diamantanedicarboxylic acid as a starting material.

6.3 Limitations of the Organic Synthesis of Diamondoids

To fabricate diamondoids—especially those with a large number of cages, that is, $n > 4$—available in larger quantities (at least grams), it is necessary to develop methods that allow synthesizing and functionalizing diamondoids more effectively than present methods. One approach that has shown to be promising is by both discharge and electric discharge plasmas generated in supercritical fluids. In the following section, we discuss the generation of pulsed laser plasmas in supercritical fluids and their application for the fabrication of nanomaterials, especially diamondoids.

The main reasons for the unsuccessful attempts for the synthesis by conventional organic synthesis are mainly:

(1) A lack of adequate precursors and catalysts
(2) The fact that with increasing number of diamondoids, the possible number of reaction pathways and reaction intermediates increases exponentially

Moreover, reaction intermediates can be trapped in local energy minima, preventing further continuation of reactions that could lead

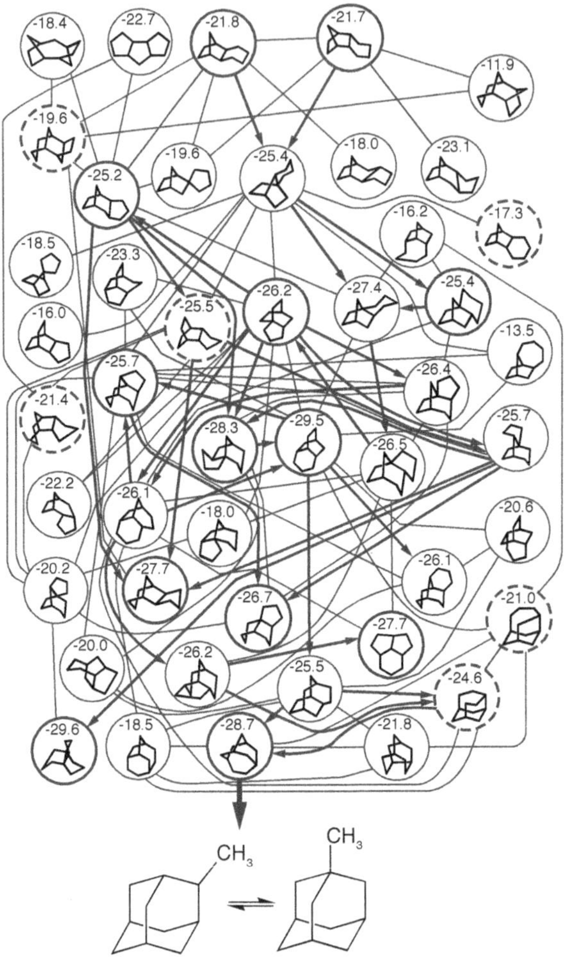

Figure 6.8 Possible reaction pathways for the synthesis of 2- and 1-methyladamantane. The figure illustrates the complexity of synthesizing diamondoids via organic synthesis. The values are indicate the standard enthalpy of formation (ΔH_f^0) in kcal.mol^{-1}. Adapted from McKervey (1980).

to the formation of a diamondoid. As an example, Fig. 6.8 displays the different reaction schemes that are possible in the synthesis of 2- and 1-methyladamantane.

PART III

NOVEL APPROACHES FOR THE SYNTHESIS OF DIAMONDOIDS BY MICROPLASMAS

Chapter 7

Diamondoid Synthesis by Electric Discharge Microplasmas in Supercritical Fluids

7.1 Introduction

In the previous chapters, we have presented a brief overview on the formation of diamondoids in natural oil and gas reservoirs and also given a summary of the different approaches to synthesize diamondoids by organic chemistry. Because of the difficulty in appropriate precursors and the increasing number of possible reaction pathways, in practice the organic synthesis of diamondoids with cage number $n \geq 4$ are not practical and so far, there have been no reports on the synthesis of diamondoids beyond tetramantane.

In this and the following chapters, we give an overview of alternative diamondoid synthesis approaches. Namely, we introduce the synthesis of diamondoids based on microplasmas, that is, using electric discharge and pulsed laser plasmas generated in supercritical fluids (SCFs) and at atmospheric pressure. Since the results presented in this and the following two chapters are based on our own work, the experimental approaches and results are

Diamondoids: Synthesis, Properties, and Applications
Sven Strauss and Kazuo Terashima

ISBN 978-981-4745-18-5 (Hardcover), 978-1-315-36466-7 (eBook)
www.panstanford.com

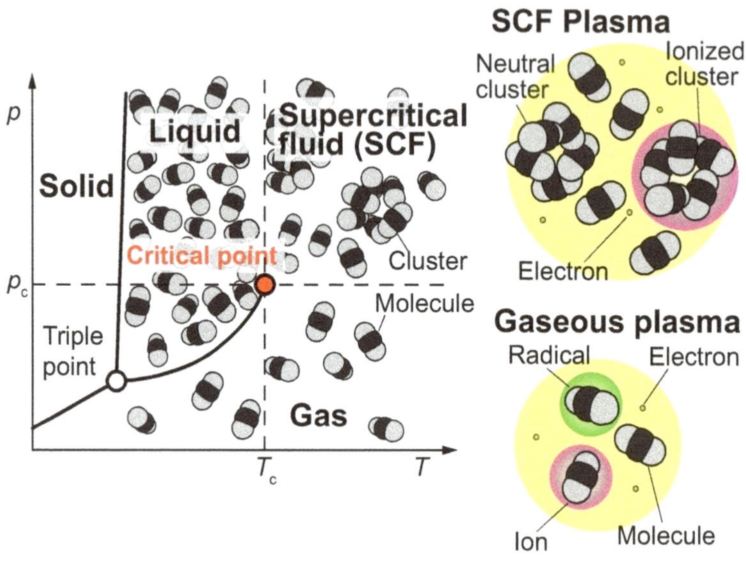

Figure 7.1 Phase diagram of a pure substance showing the different phases of a pure substance and the critical point. Compared to conventional gaseous plasmas, plasma in SCFs are expected to contain novel excited species, that is, cluster ions and cluster radicals, that are expected to enhance chemical reactions.

described in more detail compared to the previous chapters. In this chapter, we will first briefly summarize the main properties of SCFs and also present a few applications for chemistry and materials processing.

SCFs are media in a state of temperature and pressure (T, p) above the liquid–vapor critical point (CP), that is, where $T \geq T_{\text{crit}}$ and $p \geq p_{\text{crit}}$. At such conditions, the liquid and gaseous phases are not distinct anymore, and without lowering the temperature, the liquid phase cannot be obtained by simply compressing the fluid. This is illustrated in Fig. 7.1, which shows the phase diagram and the molecular structure of a pure substance in the different phases, solid, liquid, gaseous, and supercritical. Therefore, from a macroscopic point of view, SCFs represent an intermediate state between liquids and gases. As such, SCFs possess high density, high diffusivity, high solubility, and low surface tension (Jessop and Leitner, 1999). In

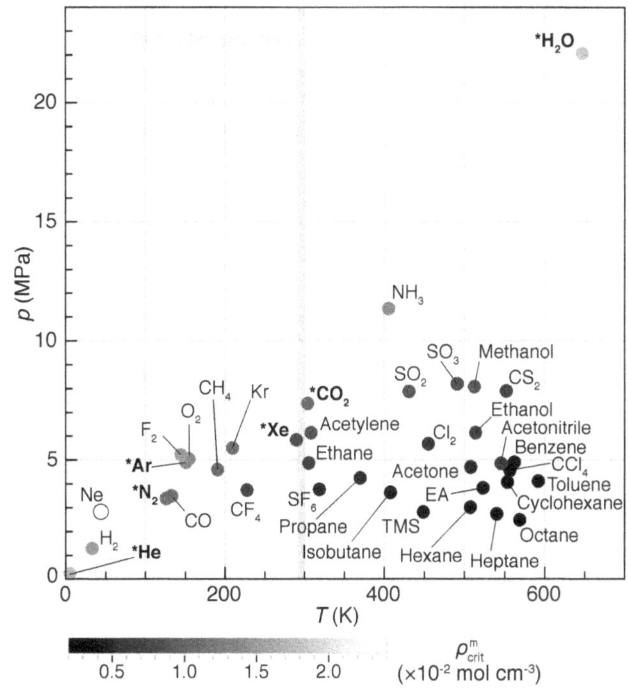

Figure 7.2 Position of critical points in relation to room temperature (between 20°C and 26°C), indicated by the vertical line. TMS, tetramethylsilane; EA, ethyl acetate. The color scale of the points indicates the molar critical density, ρ^{m}_{crit}, between 0.2 and 2.4 × e − 2 mol.cm^{-3}. The $*$ symbol marks substances inside which electric discharge plasmas have been generated in the SCF phase. Critical constants taken from Ambrose et al. (2014).

addition, their properties e.g. density, heat capacity, and dielectric constant, can be varied continuously by changing the pressure or temperature, or both.

Figure 7.2 displays the critical values (temperature, T_{crit} and pressure, p_{crit}) of various pure substances in relation to room temperature (i.e., between 20°C and 26°C), and the color scale of the CPs indicates the levels of the critical molar density, ρ^{m}_{crit}. The two extremes of T_{crit} and p_{crit} are represented by He ($T_{crit} = 5.2K$, $p_{crit} =$ 0.227 MPa), and H_2O ($T_{crit} = 647.2$ K, $p_{crit} = 22.06$ MPa). The figure also indicates the fluids that have been used for generating electric

Table 7.1 Comparison of average physical and transport properties of liquid, SCF, and gaseous states. Most of the thermophysical properties of SCFs are intermediate between those of gases and liquids. However, there are also properties that are higher compared to liquids and gases in SCFs. For example, the thermal conductivity κ reaches a maximum near the critical point

Property	Liquid	SCF	Gas
Density ($kg.m^{-3}$)	600–1600	200–900	0.6–2.0
Viscosity ($\times 10^5$ $kg.m^{-1}.s^{-1}$)	20–300	1–9	1–3
Diffusion coefficient ($\times 10^8$ $m^2.s^{-1}$)	0.02–0.2	1–40	1000–4000
Thermal conductivity ($\times 10^3$ $W.m^{-1}.K^{-1}$)	80	20–250	4–30
Surface tension ($dyne.cm^{-1}$)	20–450	0	0

discharge plasmas. So far, electric discharges have been generated inside the SCF phases of He, Ar, N_2, Xe, CO_2, and H_2O (marked by $*$ in Fig. 7.2).

Table 7.1 summarizes typical properties of liquid, supercritical, and gaseous phases. As mentioned briefly in Section 9.1 and as shown in the table, the properties of SCFs lie between those of liquids and gases.

While many properties of SCFs are intermediate between those of liquids and gases, this is not true in all cases. In the vicinity of the critical point, the thermal conductivity, specific heat, and compressibility attain a maximum, their values being significantly higher compared to the respective values of the gaseous or liquid states (Chen et al., 1999; Pittman et al., 1982; Vesovic et al., 1990; Walas, 1985).

These unique characteristics including the critical anomalies near the CP derive from the microscopic fluid structure, that is, molecular clustering of SCFs.

In the next section, we give a brief overview of the techniques necessary to generate plasmas in high-pressure media and SCFs.

7.2 Generation of Plasmas in Supercritical Fluids

The creation of plasma reaction fields in SCFs can be achieved both by electric discharges and pulsed laser ablation plasmas, thereby combining the high reactivity of plasmas with SCF-specific characteristics. In order to enable to use of plasmas in SCFs for other applications such as materials processing the control of and suppression of the gas temperature rise due to plasma generation is of great importance.

A more general overview of SCFs, the history of their discovery, experimental equipment, their main properties and applications for chemical engineering is given in the book of Jessop (Jessop and Leitner, 1999). In addition to this work, detailed descriptions of the molecular properties of SCFs have also been compiled in a separate monograph (Arai et al., 2001).

To generate electric discharges in SCFs, a high voltage (typically a few hundred volts to a few kilovolts) has to be applied between two electrodes. Applied voltage can either be a direct current (DC) or alternating current (AC).

So far, different types of electric discharges have been generated in high-pressure and supercritical fluids. For example, the generation of dielectric barrier discharge (DBD) plasmas in supercritical argon (in the pressure range between 2 to 4.9 MPa) and using at radio frequencies has been demonstrated (Suga et al., 2010). DBDs where at least one electrode is covered by a dielectric and effectively acts as a current limiter are sustained by the repeated formation of streamers, which evolve on the order of nanoseconds (Raizer, 1991). The use of DBDs has permitted the generation, even under high pressure, of nonequilibrium, low-temperature plasmas where the temperature increase is inhibited (Becker et al., 2004; Iza et al., 2008; Kogelschatz, 2007; Xu, 2001).

In this section, we introduce research related to the generation of alternating current (AC) DBDs, that facilitated the generation of low-temperature plasmas under high-pressure, including supercritical, conditions. As has been briefly mentioned in the introduction, there are many different possible configurations of DBD electrodes. In the case of high-pressure and supercritical conditions, besides a small gap distance between electrodes, an increase in the electric field, for instance by using a very pointed electrode, permits to reduce the breakdown voltage. Figure 7.3a shows a schematic of a needle-type electrode that was used for generating DBDs in high-pressure and supercritical conditions (Tomai et al., 2006). As high-

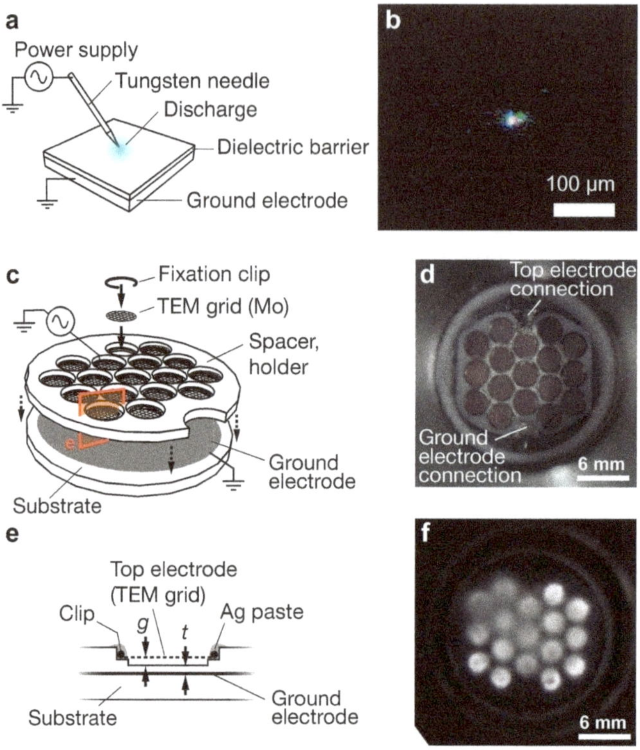

Figure 7.3 Schematics and photographs of DBD electrodes during generation. (a, b) Needle-type DBD (Kikuchi et al., 2008; Tomai et al., 2006). (c–f) Schematics of the electrodes and photographs of arrayed mesh-type electrode. Parts of the figure are adapted from Tomai et al. (2007) and Stauss et al. (2010).

voltage electrodes, the authors used Tungsten (W) needles with outer diameters of 250 μm that were fabricated by electrochemical etching. Soda-lime glass slides with thicknesses of 120–170 μm and whose back surfaces had been coated with silver paste, were used as ground electrodes. The authors generated plasmas in $scCO_2$, scN_2, scAr and scXe by applying a high AC voltage (frequency: 3 kHz; voltage: 10.5 $kV_{p\text{-}p}$) to the W needles. This electrode configuration enabled the realization of stable discharges with durations of over an hour (Kikuchi et al., 2008; Tomai et al., 2006).

Typical parameters of such DBD plasmas generated using the needle electrode geometry shown in Fig. 7.3a are summarized in Table 7.2. Current–voltage waveform measurements indicated that streamers with durations of approximately 2 ns are generated repeatedly from a few to several tens of times per cycle. While the ionization degree in these streamers is on the order of 8×10^{-4}, close to that of streamers in atmospheric-pressure DBDs (10^{-4}–10^{-3}), the high neutral density of the SCF leads to the formation of plasmas with high electron densities of $n_{\mathrm{el}} \approx 5 \times 10^{18}\ \mathrm{cm}^{-3}$.

Compared to a typical low-temperature plasma, the electron density is more than 1000 times higher. Consequently, they can be considered to be a class of particularly reactive plasmas (Tomai et al., 2006, 2007).

However, in order to increase the quantities of materials synthesized, it is necessary to enlarge the size of the plasma. One way to achieve this and without sacrificing the small gap distance is to array a number of smaller discharges. To that end, micromachined DBD electrodes have been developed (Stauss et al., 2010). The authors used commercially available TEM molybdenum mesh grids, which served as high-voltage electrodes. Machinable ceramic, serving as both a holder for the TEM grids and as dielectric barriers, were used, a schematic of the electrode geometry being shown in Fig. 7.3c. By arranging the TEM grids into a two-dimensional array, large-area-gap electrodes with a surface area of about 85 mm^2 could be realized. A photograph of such a large-area DBD generated in scXe is shown in Fig. 7.3f. In addition to the size, which in principle could be extended to even larger sizes, the advantage of using micromachined electrodes in comparison to the needle electrodes is a better control of the gap distance. With this type of electrodes, the authors succeed in obtaining stable discharges for durations of up to three hours (Stauss et al., 2010).

In another study, radio-frequency (RF) plasmas in scCO_2 were generated using a high-frequency torch at 27.12 MHz and input powers of 30 W, using electrodes with a gap distance of 1 mm (Kawashima et al., 2007). These RF plasmas using a point-to-point electrode geometry were used for the synthesis of tungsten oxide nanowires.

Table 7.2 Typical experimental conditions for needle-type DBD generated in SCF conditions (Tomai et al., 2006, 2007)

Condition	Value
Duration	2 ns
Streamer radius	< 100 μm
Repetition rate	Several to several tens / cycle
Peak current	0.1–1 mA
Current density	1.3–13 $A.cm^{-2}$
Charge density	$> 5 \times 10^{18}$ cm^{-3}
Ionization degree	$> 8 \times 10^{-4}$

Table 7.3 Comparison of experimental conditions for SCF plasma and conventional SCF processes used for synthesizing CNTs

Condition	SCF plasma synthesis	SCF synthesis
Environmental temperature (K)	304.2	1273 (Motiei et al., 2001) 823 (Lou et al., 2003)
Environmental pressure (MPa)	7.4	1000 (Motiei et al., 2001) 70 (Lou et al., 2003)
Catalyst	—	Mg (Motiei et al., 2001) Li (Lou et al., 2003)

The characteristics of the DBD plasma generated using the needle electrode shown in Fig. 7.3a are summarized in Table 7.2, and Table 7.3 lists the differences in the processing parameters between a plasma generated in an SCF and conventional processes for the synthesis of CNTs.

7.3 Electric Discharges in High-Pressure and Supercritical Fluid Microreactors

The generation of electric discharges in high-pressure and supercritical environments is not very straightforward, the high-pressure (and high-temperature) conditions making it necessary to use specialized reactor vessels adapted to such harsh conditions. From the point of view of electric discharges generated in high-pressure and supercritical media, an alternative approach, which relies on the use of microreactor systems, has been proposed.

In fact, microfluidic reactor systems have shown promise as alternative approaches to large-scale batch reactors, both in analytical chemistry and materials synthesis (McMullen and Jensen, 2010). The small volumes of microreactors allow faster mixing and more rapid heat exchange, which can lead to higher reaction yields (Hartman and Jensen, 2009), enabling the realization of ultrafast reactions (Yoshida et al., 2008), and giving unprecedented control over chemical reactions (Mason et al., 2007). In addition, they allow for more flexibility, improved safety, and less chemical waste (Mason et al., 2007). Continuous flow operation also brings about higher reproducibility and allows more accurate control of reaction times and particle sizes by adjusting the residence time (Marre and Jensen, 2010). Because of the small reaction volumes involved, microreactors can also be operated under conditions that are not easily achievable with large-scale reactors: high temperatures (Tiggelaar et al., 2005) and pressures (Verboom, 2009), or even using SCFs (Marre et al., 2010, 2008; Trachsel et al., 2009).

However, despite the advantages of microreactors over macroscale reactors, the parameters that can be changed remain limited to pressure, temperature, the concentrations of the reactants and their flow rates.

In the following, we briefly present the approach that has been adopted for the synthesis of diamondoids using electric discharges generated inside microscale reactors at high-pressure and supercritical conditions.

Figure 7.4a shows a schematic of the experimental setup (Oshima et al., 2012). The reactor consisted of fused silica capillaries with nominal inner and outer diameters of 700 ± 10 and 850 ± 20 μm, respectively, which served as fluidic channels and dielectric barriers. The electrode geometry consists of two electrodes separated by the dielectric barrier. In this case, the outer, ground electrode consisted of an indium tin oxide (ITO) thin film. The high-voltage electrode consisted of a tungsten filament (nominal diameter of 500 μm), which was inserted into the capillary.

As has already been mentioned, in the case of discharges generated in high-pressure and supercritical fluids, it is essential to control the conditions of the fluid—temperature and pressure. For the microcapillary reactors, the fluid temperature was adjusted by enclosing the capillary, and by adjusting the temperature of this shield, at temperatures of 279.2–284.8 K, which corresponds to the range of $T/T_{\text{crit}} = 0.964$–$0.983$ for xenon ($T_{\text{crit}} = 289.75$ K, $p_{\text{crit}} = 5.84$ MPa). The temperature was monitored using a thermocouple attached to the capillary.

Here too, small amount of adamantane (24.8 $\pm$ 0.4 mg were introduced into the fluid to serve as a precursor, before introducing pressurized xenon (5.83–5.99 MPa; $p/p_{\text{crit}} = 0.998$–$1.026$) into the cell. For adjusting the pressure values, syringe pumps are often used, because they allow to control the pressure and at the same also enable achieving flow rates compatible with microreactor systems. In this special case, the continuous-flow-mode operation of the syringe pump, flow rates of 0.05, 0.1, and 0.5 ml min^{-1} were realized, for which the reaction times were 100, 50, and 10 min, respectively.

The discharges were generated using sinusoidal HV signal. To generate the plasma, a sine signal wave with a frequency of 10 kHz and a voltage of 3.95–6.02 kV$_{\text{p-p}}$ (peak-to-peak) was applied (Fig. 7.4a). The detailed experimental conditions to generate the electric discharges are summarized in Table 7.4.

Gas chromatography–mass spectrometry (GC-MS) measurements were performed for further analysis of the collected material.

The power consumption of these types of discharge was found to be of the order of orders of 100 mW to 1 W (depending on the experimental setup, that is, the length of the electrodes and the gap distance of the inner electrode, the power consumption can vary).

Raman spectra acquired on solid particles indicated that the products were a mixture of sp^2- and sp^3-hybridized carbon nanomaterials, including diamondoids.

GC-MS measurements were conducted with the following three goals: 1) to estimate the amount of synthesized diamantane by integrating the peaks determined in SIM mode, 2) to determine the types of higher diamondoids that may have been synthesized, and 3) to investigate the presence of intermediate products in order to determine possible reaction mechanisms.

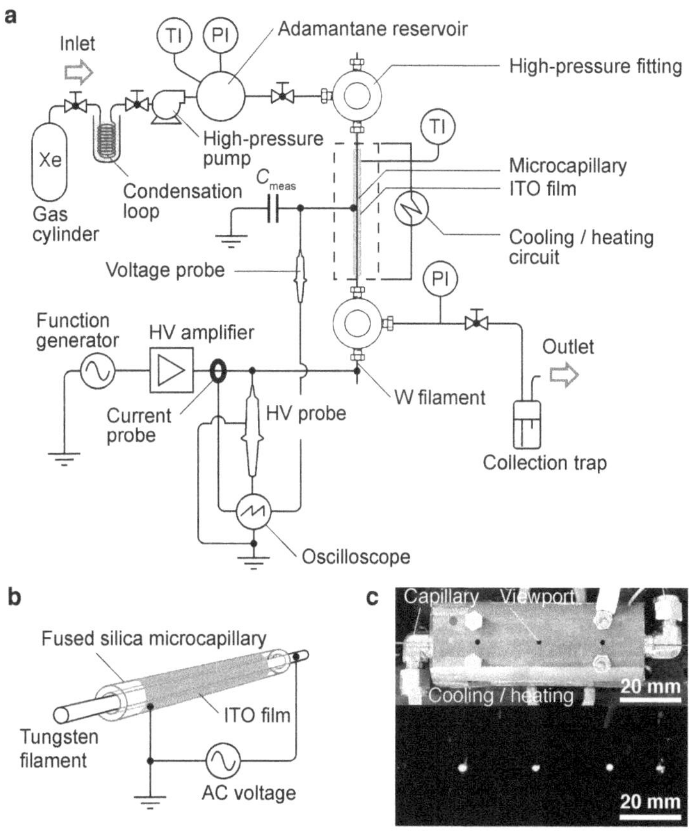

Figure 7.4 Schematic of experimental setup of high-pressure plasma microreactor. (a) Process flow diagram. The cooling loop is used to liquefy/solidify xenon gas and the high-pressure pump is used to control the flow rate. The pressure and temperature of the microreactor system were monitored using pressure and temperature indicators (PI, TI), respectively. The precursor, adamantane, was dissolved in a high-pressure cell (inner volume 3 mL) and then flowed through the reactor. I–V and Q–V curves were acquired concurrently using current and high-voltage probes connected to an oscilloscope. For Q–V measurements, a capacitor, C_{meas}, was used. (b) Detailed view of electrode geometry. The electrode consists of a fused silica capillary coated with ITO and a tungsten wire inserted into the capillary. (c) Photograph of experimental setup (top) prior to and (bottom) during plasma generation ($p = 5.90$ MPa, $T = 290.9$ K, $f = 10$ kHz, $V_{appl} = 4.6$ kV$_{p\text{-}p}$). The three viewports at the top of the cooling/heating circuit allow optical access to the capillary and the verification of plasma generation. Figure adapted from Oshima et al. (2014).

Table 7.4 Experimental conditions for diamondoid synthesis using microcapillary reactor (Oshima et al., 2014). For both batch and continuous-flow approaches, the applied frequency was $f = 10$ kHz

Parameter	Condition
Batch type	
Temperature (°C) (T/T_{crit})	15.6–16.5 (0.997–0.999)
Pressure (MPa) (p/p_{crit})	5.89–5.99 (1.009–1.023)
Applied voltage (peak-to-peak; $kV_{p\text{-}p}$)	4.32–4.78
Reaction time (min)	15–240
Adamantane (mg)	24.5–25.1
Flow type	
Temperature (°C) (T/T_{crit})	15.6–16.5 (0.997–0.999)
Pressure (MPa) (p/p_{crit})	5.83–5.96 (0.998–1.021)
Gas flow rate (mL min^{-1})	0.01, 0.05[c], 0.1[b], 0.5[a]
Peak-to-peak applied voltage ($kV_{p\text{-}p}$)	3.95–6.02
Reaction time flow (min)	10[a], 50[b], 100[c]
Adamantane (mg)	25.1–25.2

[a] Flow rate of 0.05 mL/min.
[b] Flow rate of 0.1 mL/min.
[c] Flow rate of 0.5 mL/min.
Flow rates and respective reaction times correspond to residence times of $t_{res} = 4.2, 21,$ and 42 s, respectively.

To begin with, we investigated the synthesis of diamantane ($n = 2$, $C_{14}H_{20}$, relative molecular mass $M_r = 188$). The SIM spectrum of a cyclohexane solution containing commercially available diamantane (purity > 98.0%; Tokyo Chemical Industry) with a mass-to-charge ratio (m/z) of 188 exhibited large peaks at a GC retention time of $t_{ret} = 8.3$ min.

The SIM spectra for products with m/z 188 showed peaks at the same GC retention time as the standard sample containing diamantane. Moreover, the mass spectra of the products and the standard sample at this retention time of 8.3 min exhibited similar tendencies (Fig. 7.5). The large peak at m/z 188 was attributed to the molecular ion of diamantane, and the smaller peaks at the same m/z value and intensity ratio in both spectra at lower m/z values corresponded to fragment ion peaks. On the basis of the agreement of the retention time and mass spectrum of the synthesized products with those of the standard sample, it is concluded that the synthesized products contain diamantane.

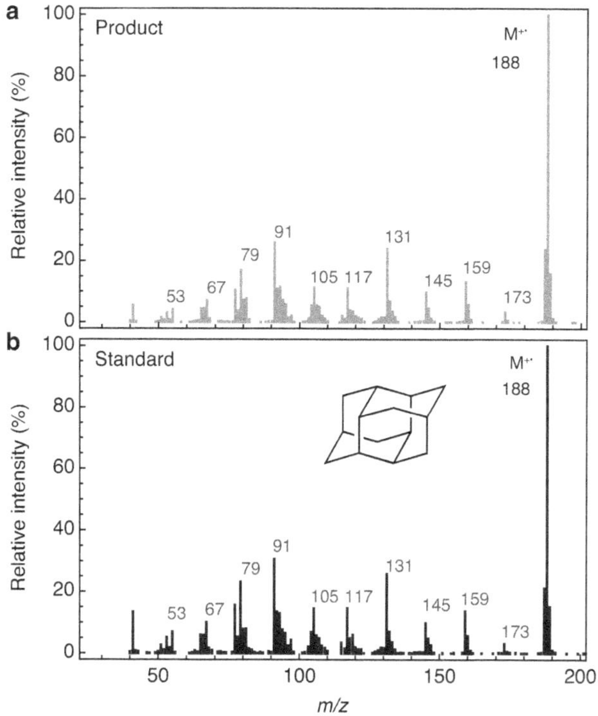

Figure 7.5 Typical mass spectra of diamantane. (a) Product synthesized in the microcapillary microplasma reactor. (b) Mass spectrum of a standard sample containing diamantane. The inset shows the molecular structure of diamantane (for clarity, hydrogen terminations are not shown). Both mass spectra were acquired at a retention time of $t_{ret} = 8.3$ min. Further, both have a molecular ion peak at m/z 188, and the structure and intensity of the fragment ion peaks of both spectra are in very good agreement. Data adapted from Oshima et al. (2014).

The quantities of diamantane were evaluated by integrating the m/z 188 SIM peak area near the retention time of 8.3 min. The area intensities were then compared with those of a calibration curve obtained by linear regression of the peak intensities measured for standard diamantane solutions with known concentrations between 10 and $1 \times 10^4 \mu g\ L^{-1}$. The maximum quantity of synthesized diamantane was found to be about ~15 μg, and the yield for the batch experiments was found to be ~600 ppm.

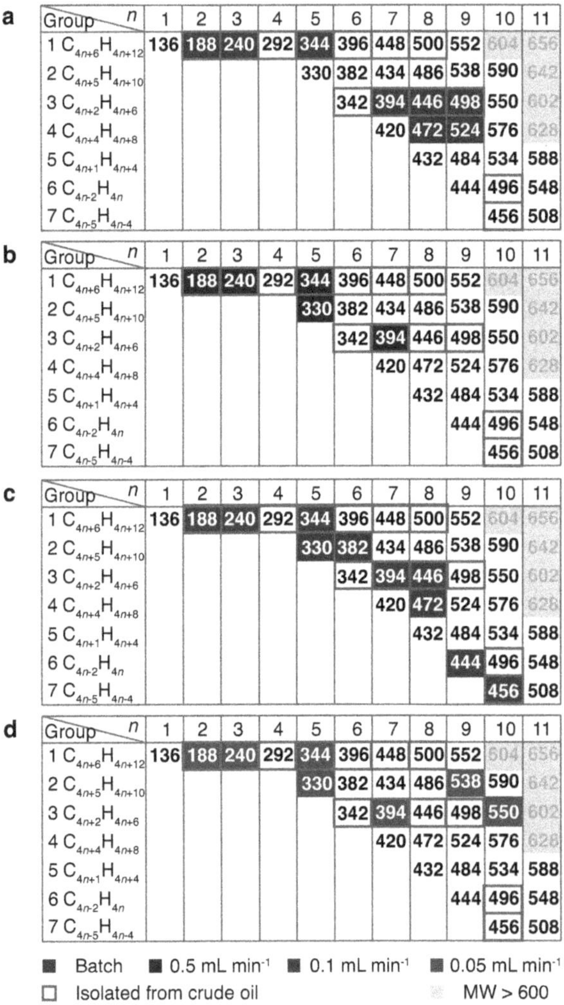

Figure 7.6 Relative molecular masses (M_r) of possibly synthesized diamondoids. Columns with $n = 1$–11 indicate the number of diamondoid cages, while rows indicate the diamondoid group. Shaded cells containing numbers in bold indicate diamondoids detected in the GC-MS measurements. (a) Batch type, (b) continuous-flow experiment with flow rates of 0.5, (c) 0.1, and (d) 0.05 mL min^{-1} (cf. experimental conditions listed in Table 7.4). Entries with orange borders correspond to diamondoids reported to have been isolated from crude oil. Cells shaded in gray in the table indicate diamondoids with $M_r > 600$; these were not investigated but are listed for completeness. Data adapted from Oshima et al. (2014).

In addition to the GC-MS spectrum of diamantane, the spectra of other diamondoids with $n > 2$ were also analyzed. Figures 7.6a–7.6d show a list of diamondoids with M_r values below 600. The columns indicate the order or cage number (n) of the diamondoids, while the rows (1–7) indicate the diamondoid group.

The easiest way to increase the order of a given diamondoid is to add an isobutyl framework (four carbon atoms and four hydrogen atoms) to any plane consisting of a cyclohexane ring in the diamondoids. This addition increases M_r by 52. For diamondoids belonging to the first group, M_r increases with $136 + 52\ (n - 1)$. On the other hand, by suitably sharing cyclohexane planes with other cages, smaller hydrocarbon units can be added to increase the order of the diamondoids. For example, C_3H_2 units are required to produce pentamantane with $M_r = 330$ from tetramantane ($M_r = 292$). The molecules on the far left in the second to seventh groups are base molecules that are obtained by adding small units to lower diamondoids, and a new cage is then grown on the left by adding C_4H_4 units.

The synthesis of triamantane and higher diamondoids was examined on the basis of other GC-MS selected ion monitoring (SIM) mode spectra. Because these diamondoids are not commercially available, it was not possible to directly compare their retention time and mass spectra with those of standard samples, as was done in the case of diamantane. However, it has been reported that the GC-MS spectra of diamondoids have two characteristics, both the retention time and mass spectra (Balaban et al., 2007; Wingert, 1992). First, the retention times of diamondoids increase almost linearly with the relative molecular mass, while those of alkylated diamondoids do not lie on a straight line.

Second, the mass spectra of higher diamondoids are nearly featureless except for strong molecular ion base peaks and peaks at $M^+/2$, which correspond to divalent molecular ions. On the basis of these features, it was determined that higher diamondoids were synthesized when the following three criteria were satisfied: (i) the SIM spectrum had a peak whose signal-to-noise (S/N) ratio was more than three; (ii) the retention time for the peak could be roughly fit on a straight line; and (iii) the mass spectrum for the retention time had a local maximum at a value that directly corresponded

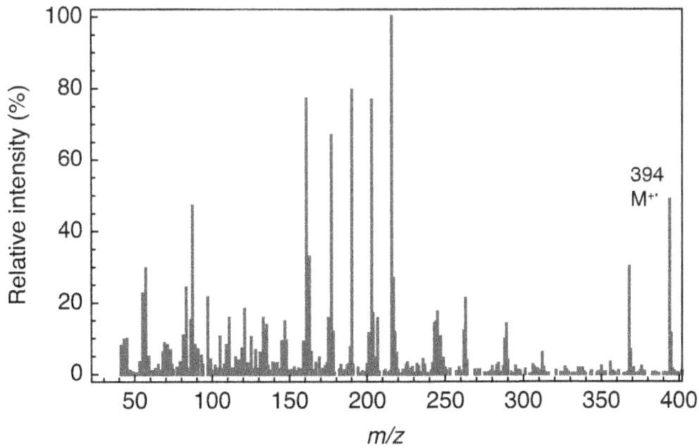

Figure 7.7 Mass spectrum of a possibly synthesized heptamantane. For this diamondoid, the number of cages is $n = 7$ and the relative molecular mass is $M_r = 394$. Data adapted from Oshima et al. (2014).

to a possible relative molecular mass of a diamondoid (as listed in Fig. 7.6). Finally, spectra that contained any peaks at m/z values higher than the expected M_r values of the diamondoids were not taken into account.

As an example, Fig. 7.7 shows a mass spectrum at a GC retention time of 22.4 min that might correspond to heptamantane ($n = 7$, $C_{30}H_{34}$, $M_r = 394$). Figure 7.6 lists the M_r values of diamondoids that may have been detected in this study. For comparison, the figure also indicates diamondoids reported to have been isolated from crude oil (Dahl et al., 2003a,b), which are the entries in Fig. 7.6a–d surrounded by orange borders.

In the batch-type experiment (cf. Fig. 7.6a), the possibly synthesized diamondoids (except diamantane) were triamantane ($n = 3$, $C_{18}H_{24}$, $M_r = 240$), one pentamantane ($n = 5$, $C_{26}H_{28}$, $M_r = 344$), one heptamantane ($n = 7$, $C_{30}H_{34}$, $M_r = 394$), two octamantanes ($n = 8$, $C_{34}H_{38}$, $M_r = 446$ and $C_{36}H_{40}$, $M_r = 472$), and two nonamantanes ($n = 9$, $C_{38}H_{42}$, $M_r = 498$ and $C_{40}H_{44}$, $M_r = 524$). Assuming that the order of diamondoids increases in steps of one inside SCF plasmas, it is peculiar that tetramantane and hexamantane were not detected.

One possible reason for this might be the difference in the dissociation energy of the diamondoids. The energy associated with the dissociation of diamondoids into lower diamondoids and hydrocarbon fragments has been calculated from ab initio simulations (Lu et al., 2005). Arranging these energies in the order of increasing n, those of hexamantane ($n = 6$) and heptamantane ($n = 7$) show local energy maxima.

It is expected that reactions in SCF plasmas will proceed not only in a direction that leads to higher diamondoids but also in the opposite direction, resulting in their dissociation and the formation of lower diamondoids. This means that only energetically stable diamondoids can survive in the reaction field. For example, most of the tetramantanes that did not grow into pentamantanes were found to dissociate back into triamantane, while pentamantanes were unlikely to form tetramantanes. Another observation that can be made regarding the types of possibly synthesized diamondoids is that they belong to the first, third, and fourth groups (rows in Fig. 7.6), whereas diamondoids belonging to the second group were not observed. In the case of pentamantane belonging to the second group ($n = 5$, $C_{25}H_{30}$, $M_r = 330$), two hydrogen atoms are close to each other, leading to high steric hindrance (Balaban et al., 2007). Furthermore, the maximum solid angle between the carbon bonds in the cage structure of this pentamantane is more than 5° larger than that of adamantane, which is free from any strain.

The extraction of this specific pentamantane and other diamondoids in the second row from crude oil has not been reported (Dahl et al., 2003a), which means that they are unstable and unlikely to form via thermal equilibrium reactions. Moreover, these diamondoids were not observed in the batch-type reactions conducted in the present study. The types of possibly synthesized diamondoids in the three experiments conducted at different flow rates are summarized in Fig. 7.6b–d.

In the experiment with a flow rate of 0.5 $mL.min^{-1}$ summarized in Fig. 7.6b, diamantane ($n = 2$, $C_{14}H_{20}$, $M_r = 188$), triamantane ($n = 3$, $C_{18}H_{24}$, $M_r = 240$), two pentamantanes ($n = 5$, $C_{26}H_{28}$, $M_r = 344$, $C_{25}H_{30}$, $M_r = 330$), and one heptamantane ($n = 7$, $C_{30}H_{34}$, $M_r = 394$) may have been synthesized. For a flow rate of 0.01 $mL.min^{-1}$ (Fig. 7.6c), the GC-MS analysis showed the presence of diamantane ($n = 2$, $C_{14}H_{20}$, $M_r = 188$) and possibly triamantane ($n = 3$, $C_{18}H_{24}$, $M_r = 240$), two pentamantanes ($n = 5$, $C_{26}H_{28}$, $M_r = 344$, $C_{25}H_{30}$, $M_r = 330$), one hexamantane ($n = 6$, $C_{29}H_{34}$, $M_r = 382$), one heptamantane ($n = 7$, $C_{30}H_{34}$, $M_r = 394$), two octamantanes ($n = 8$, $C_{34}H_{38}$, $M_r = 446$ and $C_{36}H_{40}$, $M_r = 472$), one nonamantane ($n = 9$, $C_{34}H_{36}$, $M_r = 444$), and one decamantane ($n = 10$, $C_{35}H_{36}$, $M_r = 456$).

Finally, for the experiment conducted at a flow rate of 0.05 mL min^{-1} (Fig. 7.6d), the GC-MS analysis showed the presence of diamantane ($n = 2$, $C_{14}H_{20}$, $M_r = 188$), and the possible synthesis of triamantane ($n = 3$, $C_{18}H_{24}$, $M_r = 240$), two pentamantanes ($n = 5$, $C_{26}H_{28}$, $M_r = 344$, $C_{25}H_{30}$, $M_r = 330$), one heptamantane ($n = 7$, $C_{30}H_{34}$, $M_r = 394$), one nonamantane ($n = 9$, $C_{41}H_{46}$, $M_r = 538$), and one decamantane ($n = 10$, $C_{42}H_{46}$, $M_r = 550$).

These results show three tendencies. First, in all experiments, pentamantane ($C_{26}H_{28}$, $M_r = 344$) belonging to the first group and heptamantane ($C_{30}H_{34}$, $M_r = 394$) belonging to the third group were synthesized, both of which have relatively high dissociation energies (Lu et al., 2005).

This result supports the theory that the growth and dissociation of diamondoids occur simultaneously, and that stable diamondoids are more likely to survive in the reaction field. Second, some diamondoids belonging to the second group were synthesized in the flow experiments: pentamantane ($C_{25}H_{30}$, $M_r = 330$), hexamantane ($C_{29}H_{34}$, $M_r = 382$), and nonamantane ($C_{41}H_{46}$, $M_r = 538$). These diamondoids have steric hindrance and their isolation from crude oil has not been reported (Dahl et al., 2003a,b). In addition, they were not observed in the batch-type experiment conducted in the present study. The minimum reaction time for the batch-type experiments was 15 min, longer than the longest residence time in the flow-type reactions ($t_{res} = 4.2$, 21 and 42 s). It is suggested that dissociation reactions, which lead to more stable diamondoid structures, were inhibited by the rapid flow of the products out of the plasma reaction zone; therefore, diamondoids belonging to the second group were not dissociated. This indicates that the highly nonequilibrium reaction zone of an SCF plasma allows the synthesis of molecules that cannot be formed via thermal equilibrium reactions.

Finally, as can be seen in Fig. 7.6b–d, diamondoids of increasingly high molecular mass form with increasing residence time. Focusing on diamondoids in the second group, only pentamantane ($C_{25}H_{30}$, $M_r = 330$) was observed in experiment the experiment with a flow rate of 0.5 mL.min^{-1}, which had the shortest residence time (4.2*s*). In contrast, pentamantane ($C_{25}H_{30}$, $M_r = 330$) and hexamantane ($C_{29}H_{34}$, $M_r = 382$) were observed in the experiment with a flow rate of 0.1 mL.min^{-1} ($t_{res} = 21$ s). Moreover, pentamantane ($C_{25}H_{30}$, $M_r = 330$) and nonamantane ($C_{41}H_{46}$, $M_r = 538$) were observed in the experiment with the longest residence time of 42 s (experiment with a flow rate of 0.05 mL.min^{-1}).

It appears that pentamantane ($C_{25}H_{30}$, $M_r = 330$) grows into larger diamondoids belonging to the second group. The same trend can also be observed for the diamondoids belonging to the third group. Heptamantane ($C_{30}H_{34}$, $M_r = 394$) was observed in all the experiments; octamantane ($C_{34}H_{38}$, $M_r = 446$) was obtained in the experiment with a flow rate of 0.1 mL.min^{-1}; and decamantane ($C_{42}H_{46}$, $M_r = 550$) was detected in the experiment where a flow rate of 0.05 mL.min^{-1} had been adapted. It seems natural that longer residence times lead to the synthesis of diamondoids with more cages. However, it is strange that specific diamondoids belonging to one group were observed while those belonging to other groups were not. For example, hexamantane, heptamantane, and octamantane belonging to the second group were not observed in the experiment with the highest flow rate, 0.5 mL.min^{-1}. One possible reason for this is that such diamondoids might have been produced with very low yields that were below the detection limit of GC-MS and therefore were not detected.

7.3.1 Investigation of Possible Diamondoid Reaction Paths by GC-MS Analysis of Intermediate Products

The molecular structure of adamantane suggests that the fusion of two such units does not directly lead to the formation of diamantane but to that of 1-(1-adamantyl)adamantane. To obtain diamantane using adamantane as a starting material, smaller carbon fragments have to be added to the adamantane framework. During diamond film growth by plasma CVD, hydrocarbon radicals mainly

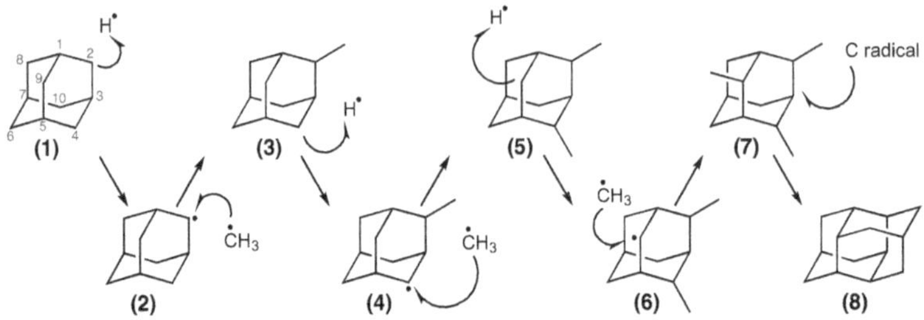

Figure 7.8 Schematic of a possible reaction mechanism for the synthesis of diamantane from adamantane. The subsequent repeated removal of hydrogen and the addition of methyl groups from adamantane **(1)** led to the formation of 2-methyladamantane **(3)**, 2,4-dimethyladamantane **(5)**, 2,4,9-trimethyladamantane **(7)**, and finally diamantane **(8)**. Numbers 1–9 in **(1)** indicate carbon sites for hydrogen removal and methyl attachment. Dots in the intermediate structures **(2)**, **(4)**, and **(6)**, and the hydrogen and methyl radicals symbolize unpaired electrons.

consisting of a single carbon atom are thought to be the main growth source (Harris, 1990). Moreover, it has been reported that the PLA of C_{60} fullerene targets containing graphite leads to the growth of larger fullerenes (Dunk et al., 2012). In that report, the authors stated that the mechanism is closed network growth, in which carbon atoms and C_2 dimers are incorporated in the fullerene network without any cleavage of the cage network. Further, the synthesis of several higher diamondoids from lower diamondoids by free-radical reactions has been reported (Dahl et al., 2010). The authors proposed the repeated removal of hydrogen atoms and the addition of methyl radicals to dangling bonds as a possible mechanism. Figure 7.8 illustrates such a possible reaction mechanism for the synthesis of diamantane from adamantane.

This type of reaction may also be valid in for the synthesis of diamondoids using electric discharges in supercritical xenon. To confirm this hypothesis, the presence of methyl derivatives of diamondoids was investigated in the GC-MS scan mode. First, SIM chromatograms for m/z 150, which corresponds to the relative molecular mass of methyladamantane, exhibited peaks at a GC retention time of 5.1 min. As shown in Fig. 7.9, the mass spectrum acquired at 5.1 min (Fig. 7.9a) is in good agreement with that of

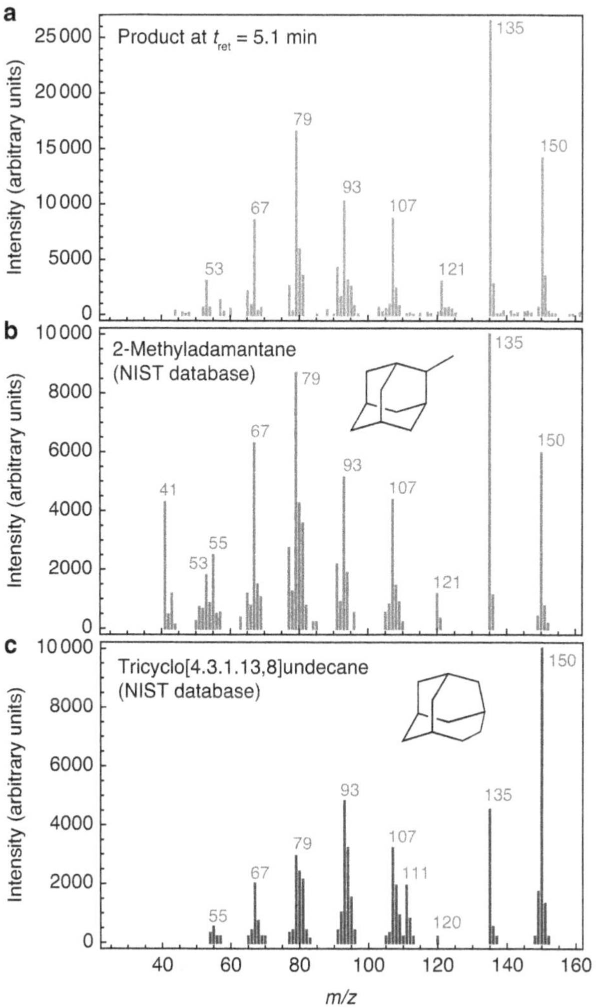

Figure 7.9 Mass spectra of synthesized products close to the retention time of adamantane. (a) Mass spectrum acquired for $t_{ret} = 5.1$ min. (b) Reference mass spectrum and molecular structure of 2-methyladamantane. (c) Reference mass spectrum of tricyclo[4.3.1.13,8]undecane (homoadamantane). As can be seen in the molecular structure of homoadamantane, two of the rings consist of six carbon atoms, while the other rings consist of seven. All spectra have a molecular ion peak at m/z 150; however, the characteristic fragment ion peaks and their respective intensities indicate the formation of 2-methyladamantane. Data adapted from Oshima et al. (2014).

2-methyladamantane (NIST/EPA/NIH Mass Spectral Library (EI), 2012), which is presented in Fig. 7.9b.

On the other hand, the mass spectrum of homoadamantane, whose relative molecular mass is 150 but which has one additional carbon atom introduced into one of the cycles in the adamantane cage, shows the highest intensity peak at m/z 150. The largest peak at m/z 135 suggests that a methyl unit can easily dissociate from the adamantane framework during the electron ionization process in the mass spectrometer. This result suggests that the first step in producing diamantane from adamantane is the substitution of a hydrogen atom of adamantane by a methyl radical. However, from GC-MS measurements alone, it is not possible to infer which carbon atom, secondary or tertiary, loses its hydrogen atom.

The presence of dimethyl- and trimethyladamantanes, whose relative molecular masses are 164 and 178, respectively, were also investigated. The chromatograms contained peaks that correspond to these m/z values, their mass spectra exhibiting molecular ion peaks (Fig. 7.10) and larger peaks at m/z values 15 less than those of the respective molecular ion peaks. Other possible derivatives of adamantane with M_r values of 164 and 178 are ethyl- and propyladamantanes, respectively. However, the mass spectra of these two derivatives show peaks at m/z 135 (NIST/EPA/NIH Mass Spectral Library (EI), 2012), because the ethyl and propyl groups are easily dissociated from their bases by electron ionization. Therefore, the observed spectra are more likely to be those of dimethyl- or trimethyladamantane than ethyl- or propyladamantane.

On the basis of this GC-MS spectra analysis, it is suggested that methyl-, dimethyl-, and trimethyladamantanes are intermediate products in the synthesis of diamantane from adamantane. Considering the possible intermediate structures that appear in the formation of diamantane, the sites of the two methyl groups in dimethyladamantane must be 2 and 4, and those of trimethyladamantane must be 2, 4, and 9 (cf. numbering in structure **(1)** of Fig. 7.8). Experimentally, 1,3-dimethyladamantane cannot be distinguished from 2,4-dimethyladamantane and 1,3,5-trimethyladamantane from 2,4,9-trimethyladamantane by GC-MS. It is reasonable to assume that many different types of methyl derivatives of adamantane are formed in parallel, and only specific

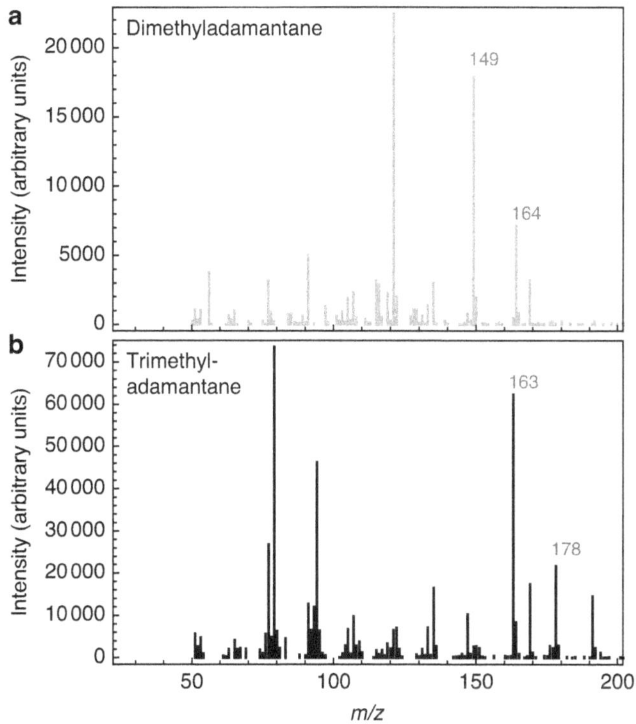

Figure 7.10 Mass spectra of reaction intermediates produced during the synthesis of diamantane. (a) Dimethyladamantane (M_r = 164) and (b) trimethyladamantane (M_r = 178). Data adapted from Oshima et al. (2014).

derivatives grow into diamantane. Although these derivatives do not lead to the formation of diamantane, mass spectra that may correspond to tetramethyl and pentamethyladamantane were observed. Moreover, mass spectra corresponding to six molecules, namely, methyl, dimethyl, and trimethyl derivatives of diamantane and triamantane, were also observed.

In summary, the proposed growth mechanism that leads to the formation of diamantane in SCF plasmas consists of the repetition of two processes: (1) the removal of hydrogen atoms and (2) the addition of methyl radicals to dangling bonds. Because the carbon source was limited to adamantane in this study, the additional methyl radicals most likely formed by the decomposition of adamantane in the plasma. In the cage closure reaction from trimethyladamantane to diamantane, other species such as carbon ions might have been formed (Tsuda et al., 1986).

To achieve higher reaction yields, the promotion of elementary steps such as the control of the electron temperature to remove dangling hydrogen bonds more effectively, and the addition of hydrocarbon sources to produce more methyl radicals is considered a promising strategy.

In this study, the GC-MS spectra were obtained from solutions collected after the synthesis and some of the gaseous products may have been lost because of evaporation. For completely understanding diamondoid growth and determining the species involved, one effective approach might be to conduct online monitoring of gaseous products by connecting the microreactor directly to a mass spectrometer (Licklider et al., 1995). Moreover, ab initio calculations may help in determining possible reaction pathways and in establishing the positions of additional methyl groups.

Diamondoids were synthesized using an SCF plasma microreactor; adamantane was used as the precursor in this synthesis. GC-MS spectra indicated that diamondoids up to decamantane ($n = 10$) may have been synthesized. On the basis of GC-MS spectra that corresponded to intermediate products, it is suggested that diamondoids grew sequentially by the substitution of hydrogen terminations with methyl radicals. Factors related to the physicochemical properties of the diamondoids, that is, dissociation energy, steric hindrance, and residence time, are considered to control the types of synthesized diamondoids. In future work, it might be possible to achieve the selective growth of higher diamondoids by controlling the radical supply to the diamondoid precursors, using various hydrocarbon additives or a combination of photochemical and plasma activation.

Finally, the present approach of using microplasmas in high-pressure and SCF microfluidic reactors may be a promising strategy for the synthesis of other nanomaterials.

Chapter 8

Synthesis of Diamondoids by Pulsed Laser Plasmas

Pulsed laser ablation (PLA) by plasmas that are generated by irradiating a solid target has been used since the appearance of Ruby lasers in the early 1960s. Because of its versatility, PLA has attracted a lot of attention, and it has shown a high potential for a wide range of materials processing: deposition of thin solid films, nanocrystal growth, surface cleaning and the fabrication of microelectronic devices.

Compared to PLA in vacuum or gaseous environments, the interactions between a laser pulse and a solid target in a confining liquid, solution, or high-pressure medium are more complex, and the fundamental aspects of PLA of solid targets in such conditions is still only poorly understood.

It is also important to note that the ablation mechanisms causing a plasma plume are different for nanopulse and pico- or femtosecond laser pulses. While the detailed mechanisms are still not understood yet, in the case of nanosecond pulses, material ejection is thought to be dominated by thermal processes (Ashfold et al., 2004). In nanosecond lasers, there are three main stages that lead to the formation of a plasma. Laser photons couple both with

Diamondoids: Synthesis, Properties, and Applications
Sven Strauss and Kazuo Terashima

ISBN 978-981-4745-18-5 (Hardcover), 978-1-315-36466-7 (eBook)
www.panstanford.com

electrons and phonons of the target material. The photon–electron coupling results in an immediate rise of the electron temperature, leading to vaporization of the target.

In the case of ultrashort (i.e., pico- or femtosecond) laser pulses, the plasma formation mechanism is different. It is generally a nonequilibrium process and the heating, forming of the plasma plume, and material ejection occur after the laser pulse (Saito et al., 2002, 2003; Sakka et al., 2002a,b). Electrons are excited only to a few or few tens of electron volts. Consequently, the lattice temperature remains unchanged, the main amount of the laser pulse energy being mainly absorbed in a thin layer close to the surface. Therefore it is more the photon absorption depth instead of the thermal diffusion depth that governs the heated volume (Ogale et al., 1987; Patil et al., 1987).

Besides the generation of pulsed laser plasmas in vacuum, PLA plasmas have recently found increased use in liquids, for both the synthesis and functionalization of nanoparticles (Amendola and Meneghetti, 2009). In addition to mainly metallic nanoparticles, PLA in liquid water has also been used for the synthesis of nanodiamonds (Amans et al., 2009). The nanodiamond particle obtained in this fashion consisted of a diamond core with diameters ranging from 5 to 15 nm, surrounded by a graphitic shell with a thickness ranging between 3 and 4 nm. It was found that the synthesis of nanodiamonds occurs in a very narrow temporal window, at the early stages of plasma formation. The shockwave duration was found to be roughly twice the pulse duration ($\sim$10 ns), during which the pressure inside the shockwave reached values between 4.5 and 22.5 GPa. After the shockwave subdues, the pressure values are too low for generating diamondoids and only sp^2 carbon is formed.

Figure 8.1 illustrates the mechanisms leading to plasma formation by irradiation of a target by a pulsed laser source, the generation of shockwaves inside the target and the fluid, and the formation and collapse of a cavitation bubble. In previous work (Saitow and Yamamura, 2009), the authors demonstrated that the size of metallic nanoparticles could be adjusted by changing the properties of the supercritical medium.

Using PLA of highly oriented pyrolytic graphite (HOPG), the synthesis of diamondoids was investigated (Nakahara et al., 2011,

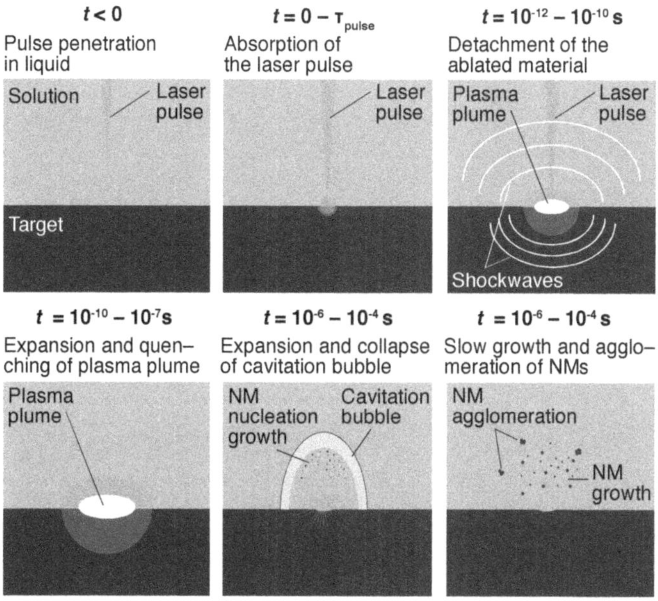

Figure 8.1 Mechanisms of plasma formation and material removal by pulsed laser ablation. The different snapshots illustrate the formation of the plasma and subsequent generation of a shockwave inside the target and the fluid, followed by the growth and collapse of a cavitation bubble, and nanomaterial (NM) nucleation, growth, and agglomeration.

2010). Xenon and CO_2 are the two gases whose critical temperature is close to room temperature. However, while dielectric barrier discharges can be generated relatively easily in supercritical xenon (scXe), the main disadvantage of Xe is its prohibitively high cost, which is about 1000 times more expensive than other supercritical fluid (SCF) media such as Ar and CO_2 (Jessop and Leitner, 1999). This makes Xe less attractive as a supercritical medium for both science and industry.

On the other hand, $scCO_2$ is one of the most widely used SCFs, having already found application in various fields, as it has been described in the brief introduction on SCFs above.

For generating plasmas in SCFs, the use of high-pressure cells that can withstand the high pressure of SCF media is necessary.

One design of a high-pressure cell that was used for realizing pulsed laser plasmas in liquids and SCFs is presented in Fig. 8.2.

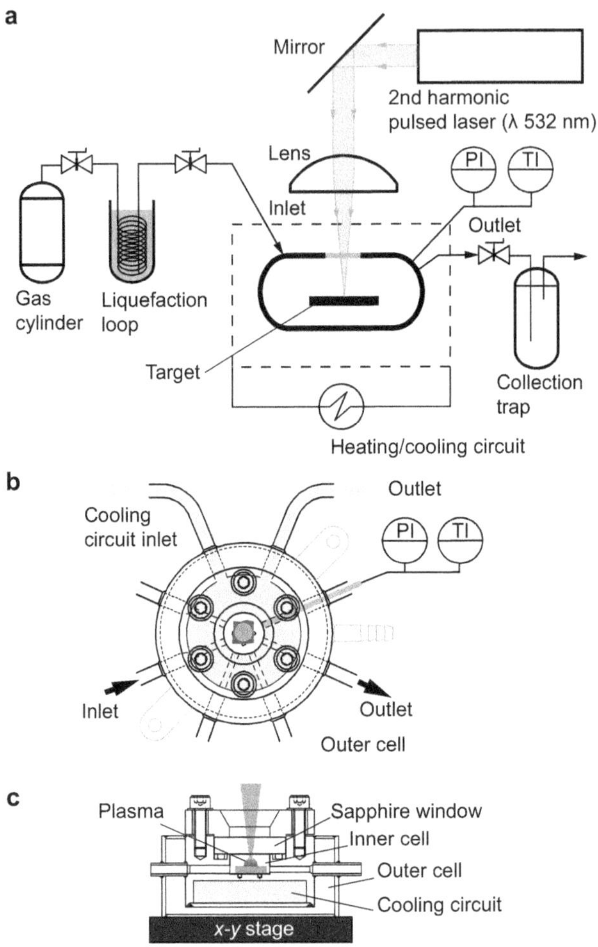

Figure 8.2 Experimental setup for realizing diamondoid synthesis by pulsed laser ablation in $scCO_2$ and scXe. (a) Schematic of the experimental setup consisting of a stainless steel high-pressure cell, a cooling/heating circuit, a compression loop for pressurizing the gas and pressure, and temperature indicators (PI and TI). At the end of the experiment, the chamber is purged and the synthesized materials collected in a trap connected to the outlet. (b) Top view of the high-pressure cell and cooling/heating circuit that allows us to adjust the temperature in the inner cell. (c) Side view of the high-pressure cell. The pulsed laser light is focused onto a target through a sapphire window and the high intensity of the laser leads to rapid heating of a zone close to the laser spot and formation of the plasma.

The main part of the equipment consists of a stainless steel cell that contains the pressurized gas. The temperature and the pressure in the inner cell can be monitored by a thermocouple and pressure sensor, respectively. The outer cell contains a water circuit that allows us to control the temperature in the inner cell.

For condensing Xe and to reach SCF conditions, the gas is flown through a liquefaction loop cooled by liquid nitrogen. In the case of CO_2, the gas is usually compressed using a high-pressure pump. The compressed gas is then introduced into the cell. After the experiment, the synthesized materials are collected in a liquid trap connected to the outlet of the high-pressure cell.

8.1 Application of Pulsed Laser Plasmas in Supercritical Fluids to Nanomaterial Synthesis

Figure 8.3 shows a series of shadowgraph images that were acquired near the critical point of CO_2. The images show the variation of the fluid after the irradiation by the laser pulse. At a delay of 3 μs after the irradiation, one can observe the formation of a shockwave. Then, after about 17 μs, one can observe the formation of transparent shell. This double structure is only observed for conditions close to the critical point.

In contrast to PLA in liquids, PLA in SCFs has not been used as much up to now. One example of PLA is the fabrication of gold nanoparticles by PLA in $scCO_2$ (Saitow and Yamamura, 2009). In addition to noble metal nanoparticles, the same group also investigated the formation of silicon nanocrystals (Si-nc) by PLA of Si targets. They could show that the size and crystal structure of the Si-nc (core vs. shell) could be changed by adjusting the density of the SCF medium. Close to the critical point, the cooling rate is maximum.

Sasaki et al. (Sasaki and Aoki, 2008) investigated the effect of the pressure of the supercritical medium on the formation of nanoparticles.

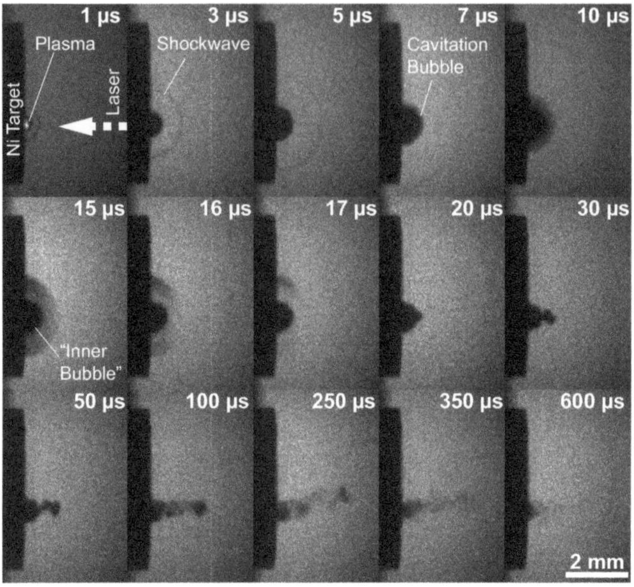

Figure 8.3 Series of shadowgraph images acquired in liquid $\mathbf{CO_2}$. The conditions of the fluid were close to the critical point ($T = 302.0\,\text{K}$, $p = 7.30\,\text{MPa}$). Between 7 and 17 μs, the cavitation bubble displays a particular structure that consists of an inner bubble and an outer, partly transparent shell. Images adapted from Urabe et al. (2013).

8.2 Synthesis of Diamondoids by Pulsed Laser Plasmas

In the following, we describe the application of PLA in supercritical CO_2 ($scCO_2$) for the synthesis of diamondoids (Nakahara et al., 2011, 2010). The authors used a 3 mL high-pressure cell (see Fig. 8.2c), and a highly oriented pyrolitic graphite (HOPG) target, which was placed in a batch-type high-pressure cell (inner volume: $3\,\text{cm}^3$). Adamantane (purity >99.0%) was dissolved in $scCO_2$ with and without cyclohexane as a co-solvent; see the experimental conditions listed in Table 8.1.

The high-density CO_2 was realized by condensation of the CO_2 gas (purity > 99.995%) in a liquefaction loop cooled by liquid nitrogen before introducing it into the inner cell. A second-

Table 8.1 Experimental conditions for PLA in $scCO_2$ with and without cyclohexane adopted in the study of Nakahara et al. (2011)

Parameter	$scCO_2$	$scCO_2$ with cyclohexane
Adamantane concentration ($mol\,L^{-1}$)	$(1.2–12) \times 10^{-2}$	$(3.1–6.2) \times 10^{-2}$
Cyclohexane concentration ($mol\,L^{-1}$)	–	7.7×10^{-1}
Temperature T (K)	304.5–305.2	304.5–304.8
Reduced temperature (T/T_{crit})	1.001–1.004	1.001–1.002
Pressure p (MPa)	7.42–7.59	7.54–7.56
Reduced pressure (p/p_{crit})	1.005–1.028	1.022–1.024

harmonic neodymium-doped yttrium aluminum garnet (Nd:YAG) laser ($\lambda = 532$ nm) was used for the ablation of the HOPG targets in both scXe (Nakahara et al., 2010) and $scCO_2$. The maximum energy was 7 mJ per pulse, giving a maximum fluence of approximately 18 J.cm^{-2} on the surface of the HOPG target. The pulse width was 7 ns, with a repetition rate of 10 Hz. The ablation experiments were conducted for periods of 60 min. The synthesized products were retrieved by evacuating the gas through a collection trap containing 10 mL of cyclohexane.

8.3 Micro-Raman Spectroscopy

Figure 8.4 is a collection of Raman spectra of products isolated after PLA in $scCO_2$ with and without cyclohexane as a cosolvent (both before and after pyrolysis) (Nakahara et al., 2011), together with the Raman spectrum of the material obtained by PLA in scXe (Nakahara et al., 2010).

For illustrating the effect of increasing cage numbers on the Raman spectra, in Fig. 8.4f,g the spectra of commercial adamantane, and [12312] hexamantane (Filik et al., 2006b) are also included. The Raman spectra of the products obtained by PLA in $scCO_2$, regardless of the presence of cyclohexane, presented almost the same features that are known to be characteristics of Raman spectra of diamondoids, that is, well-resolved peaks in the range

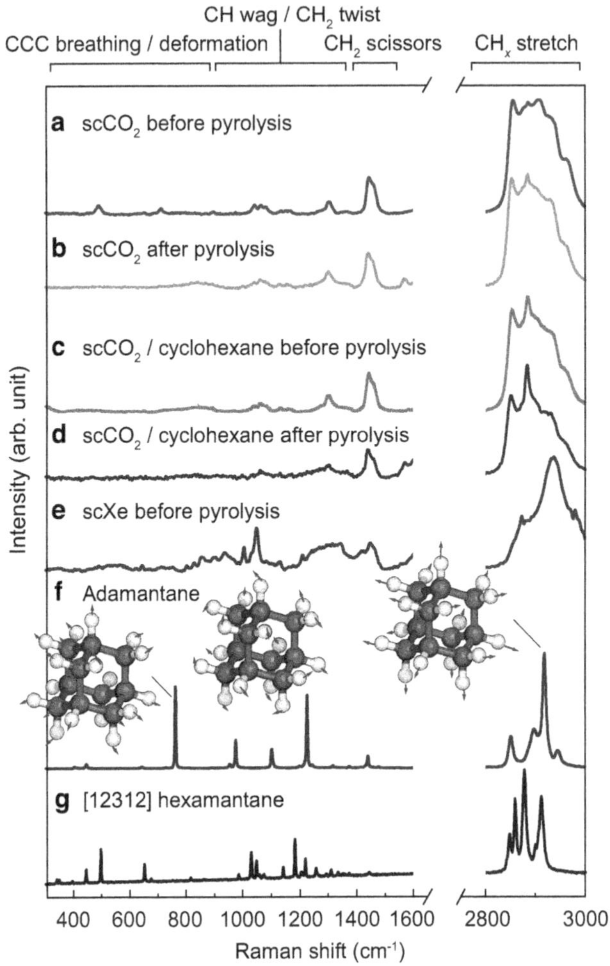

Figure 8.4 Micro-Raman spectra of synthesized products obtained by PLA and compared to spectra obtained on diamondoids isolated from crude oil. (a) Products synthesized in $scCO_2$ before pyrolysis. (b) Spectrum obtained on products after pyrolysis. (c) In $scCO_2$ with cyclohexane before pyrolysis. (d) After pyrolysis. (e) In scXe before pyrolysis (Nakahara et al., 2011). For illustrating the effect of increasing cage number, the Raman spectra of (f) commercial adamantane and (g) [12312]hexamantane (Filik et al., 2006b) are included as examples. (f) The three strongest Raman active vibration modes for adamantane, corresponding to C–C breathing, CH wag/CH_2 twist, and CH_x stretching modes, respectively, are also indicated. Data adapted from Nakahara et al. (2011).

1000–1500 cm^{-1} and strong peaks in the high-energy region, 2800–2950 cm^{-1}.

On the other hand, the peaks in the measured Raman spectra are supposed to originate from a mixture of different hydrocarbons. Moreover, it is known that the Raman scattering signal varies largely for each material, for example, sp^2-hybridized carbon shows a much stronger Raman signal compared to that of sp^3-hybridized carbon (Ballutaud et al., 2008; Wada et al., 1980). Therefore, it is possible that the peaks from synthesized diamondoids and those from other hydrocarbons such as paraffin and aromatics get balanced out, which could explain the lack of strong peaks in the low-energy region below 500 cm^{-1}. The features in the wavenumber range between 1000–1500 cm^{-1} and the peaks in the range between 2800–2950 cm^{-1}, which are characteristic of sp^3 CH_x stretching modes (Ballutaud et al., 2008), indicate the presence of diamond-structured hydrocarbons in the synthesized products. The existence of diamond-structured hydrocarbons in the synthesized products was also indicated for PLA in $scCO_2$, both with and without cyclohexane. In all the Raman spectra of the synthesized materials, no indication of the precursor adamantane could be found. This is attributed to the fact that adamantane sublimes even at room temperature and that for increasing the concentration of the samples, the solutions collected after PLA were evaporated. Interestingly, the Raman spectra collected from the products obtained by PLA in $scCO_2$ with and without cyclohexane showed similar characteristics. Thus, even when cyclohexane was added as a cosolvent, the synthesized materials, which possess relatively high Raman scattering cross sections, seemed to be almost identical for the products obtained by PLA in neat $scCO_2$. Smaller differences that are found in the Raman spectra for the samples obtained in $scCO_2$ which cyclohexane as a cosolvent are attributed to the possible synthesis of additional species which could be formed by the dissociation of cyclohexane. In contrast, the differences between the products obtained by PLA in $scCO_2$ and scXe (Nakahara et al., 2010) were bigger. For example, the strongest intensities in the low-energy region (300–1500 cm^{-1}) appeared around 1440 cm^{-1} for PLA in $scCO_2$ with and without cyclohexane, while the most intense Raman peaks appeared around 1050 cm^{-1} for PLA in scXe. Moreover, the shapes of the strong peaks in the high-energy region (2800–3000 cm^{-1}) vary considerably. These dissimilarities indicate that the synthesized materials were different for PLA in $scCO_2$ and scXe. The effect of pyrolysis for the samples obtained by PLA in $scCO_2$ and in $scCO_2$/cyclohexane are shown in Figs. 8.4a,b and 8.4c,d, respectively. Compared to the spectra before pyrolysis, the biggest difference can be observed in the high-wavenumber region between 2980 and 3000 cm^{-1}. The intensity of the peaks that are attributed to sp^2 CH_x stretching modes (Ballutaud et al., 2008) were lower after pyrolysis. In addition, the peaks in the high-wavenumber region become better resolved. From this we conclude that pyrolysis permitted to reduce the number of nondiamondoid hydrocarbons. However, we assume nondiamondoid hydrocarbons to persist in the samples even after pyrolysis and that more separation steps are necessary to separate diamondoids from nondiamondoids.

8.4 Gas Chromatography–Mass Spectrometry

8.4.1 Synthesis of Diamantane

The retention times for m/z 136 and m/z 188, which correspond to the molecular weights (MWs) of adamantane and diamantane ($n = 2$), respectively, were compared for products obtained by PLA in $scCO_2$ with and without cyclohexane, and the standard solution with dissolved commercial adamantane and diamantane. The standard sample was prepared by dissolving both commercially available adamantane and diamantane (purity >98.0%) in cyclohexane, each with a concentration of 5.0 $mg.L^{-1}$. The comparison revealed the same time lag, 3.5 min, for all experimental conditions. Moreover, the mass spectra of the gas chromatography-mass spectrometry (GC-MS) measurements at the retention times of selected ion monitoring (SIM) for m/z 188 for both products from PLA in $scCO_2$ with and without cyclohexane presented almost the same fragments as those of the standard diamantane; see Fig. 8.5.

From this data, it can be concluded that diamantane was synthesized. To estimate the production rate of diamantane, a

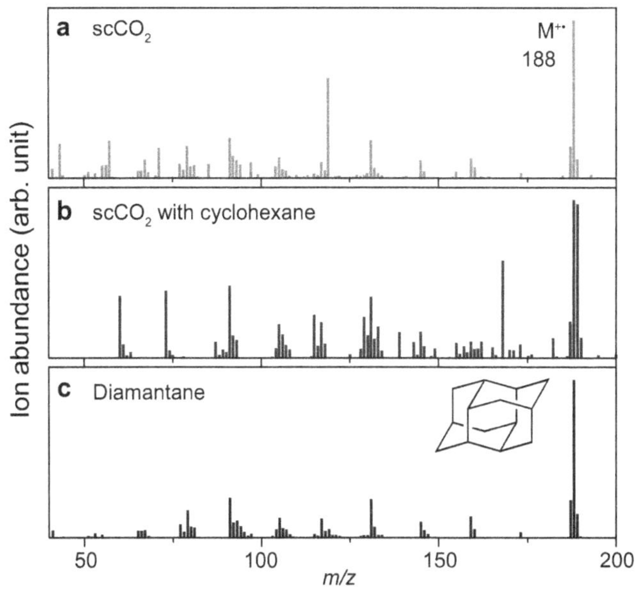

Figure 8.5 GC-MS mass spectra acquired at a retention time of $t = 7.3$ min. (a) Synthesized product obtained in neat $scCO_2$. (b) Product synthesized in $scCO_2$ and cyclohexane. (c) Mass spectrum of diamantane of the standard sample. The inset shows the molecular structure of diamantane. The main feature of the mass spectra is the molecular ion peak at m/z 188 and fragment peaks that are characteristic of diamantane. Data adapted from Nakahara et al. (2011).

calibration curve was constructed. For this, cyclohexane solutions with diamantane concentrations of 0.0, 1.0, 2.0, 4.0, and 10.0 $mg.L^{-1}$ were prepared. Figure 8.6 shows the comparison of the SIM peak areas as a function of the concentration of diamantane. By comparing the SIM peak areas of the synthesized products, the production rate of diamantane was estimated to be 0.2 μgh^{-1} for PLA in both $scCO_2$ with and without cyclohexane; see Table 8.2.

8.4.2 Possible Synthesis of Diamondoids with $n \geq 3$

SIM curves and mass spectra indicated the possible synthesis of diamondoids consisting of three or more cages by PLA in $scCO_2$. At first, however, only triamantane ($n = 3$) was observed in addition to

Table 8.2 SIM peak area and estimated diamantane concentration of the product obtained by PLA in $scCO_2$

Supercritical medium for PLA	Solution for GC-MS analyses	SIM peak area (arb. unit)	Diamantane concentration ($mg\,L^{-1}$)
$scCO_2$	Sample in 1 mL cyclohexane	2100	0.23
$scCO_2$ w. cyclohexane	Sample in 1 mL cyclohexane	1900	0.21

Data adapted from Nakahara et al. (2011).

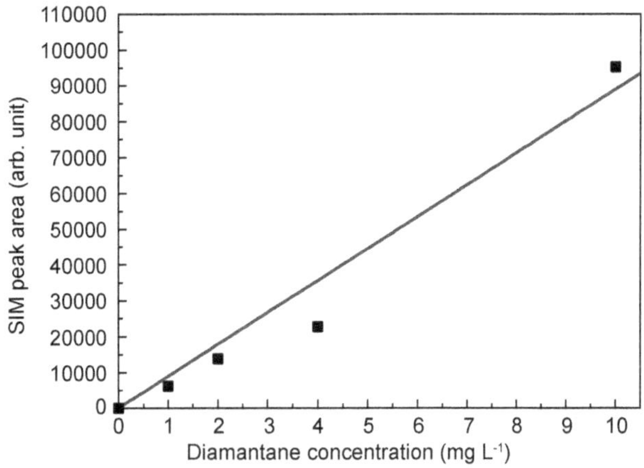

Figure 8.6 Calibration curve fitted to the GC-MS SIM peak area corresponding to m/z 188 with cyclohexane solution with dissolved diamantane. The GC-MS scan rate was 1.79 scans.s^{-1}. Data adapted from Nakahara et al. (2011).

diamantane for PLA in pure $scCO_2$. We suggest that this might be due to the poor solubility of adamantane in $scCO_2$. To improve the signal-to-noise ratio in the GC-MS measurements, the sample solution was enriched by using a smaller amount of solvent (cyclohexane of 20 μl instead of 1 mL). As a result, more peaks were observed that could be attributed to diamondoids consisting of up to 12 cages. The cage numbers of eluted diamondoids obtained by PLA in $scCO_2$ with and without cyclohexane as a function of the experimentally determined GC-MS retention times are listed in Tables 8.3 and 8.4, and Fig. 8.7, respectively.

Table 8.3 List of possibly synthesized diamondoids by PLA in $scCO_2$, with their molecular formulae, MWs, GC-MS relative retention times, and dual graph codes

Cage number (n)	Molecular formula	M^+ (m/z) base peak	GC-MS relative retention time	Structure
2	$C_{14}H_{20}$	188	1.000	Diamantane
3	$C_{18}H_{24}$	240	1.189	Triamantane
4	$C_{22}H_{28}$	292	1.509	e.g., [121]
6	$C_{29}H_{34}$	382	1.986	e.g., [12(1)31]
	$C_{30}H_{36}$	396	2.038	e.g., [12121]
7	$C_{30}H_{34}$	394	1.843	e.g., [123121]
10	$C_{35}H_{36}$	456	2.111	[1231241(2)3]
12	$C_{45}H_{46}$	586	2.495	e.g., [12131431234]

Data adapted from Nakahara et al. (2011).

Table 8.4 List of possibly synthesized diamondoids by PLA in $scCO_2$ with cyclohexane as a cosolvent, with their molecular formulae, MWs, GC-MS relative retention times, and dual graph codes

Cage number (n)	Molecular formula	M^+ (m/z) base peak	GC-MS relative retention time	Structure
2	$C_{14}H_{20}$	188	1.000	Diamantane
3	$C_{18}H_{24}$	240	1.188	Triamantane
4	$C_{22}H_{28}$	292	1.510	e.g., [121]
6	$C_{26}H_{30}$	342	1.627	[12312]
	$C_{29}H_{34}$	382	1.987	e.g., [12(1)31]
	$C_{30}H_{36}$	396	2.039	e.g., [12121]
8	$C_{34}H_{38}$	446	2.156	e.g., [1213(1)21]
10	$C_{45}H_{50}$	590	2.423	e.g., [12(3)1(2)3(1)23]
12	$C_{45}H_{46}$	586	2.493	e.g., [12131431234]

Data adapted from Nakahara et al. (2011).

The relative retention times were calculated by dividing the individual retention times by the retention time of diamantane in the range 7.148–7.157 min. The GC-MS signals can be assigned to the following diamondoids: triamantane at m/z 240; tetramantane at m/z 292; hexamantanes at m/z 342, 382, and 396; heptamantane at m/z 394; octamantane at m/z 446; decamantanes at m/z 456 ($C_{45}H_{46}$) and 590; and dodecamantane at m/z 586. From Fig. 8.7 once can see that the GC-MS retention times increase almost linearly

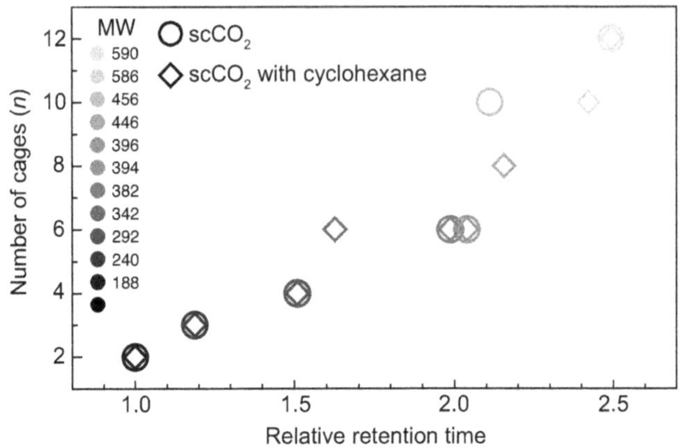

Figure 8.7 Variation of the number of diamondoid cages obtained by $\mathbf{CO_2}$ with and without cyclohexane as a function of the GC-MS relative retention time. The retention time of diamantane (t = 7.148–7.157 min) is taken as the reference retention time and the color map indicates the MW of the detected diamondoids. Diamondoids with higher MWs need increasingly longer elution times for being detected by MS, the increase being almost linear. Data adapted from Nakahara et al. (2011).

with the number of cages, which is in agreement with previous work (Balaban et al., 2007).

Furthermore, the relative retention times of the diamondoids obtained by PLA in $scCO_2$ both with and without cyclohexane showed good agreement: triamantane (1.188–1.189); tetramantane (1.509–1.510); hexamantanes (1.986–1.987 for a MW of 382 and 2.038–2.039 for a MW of 396); and dodecamantane (2.493–2.495). In addition, some of the observed diamondoids, corresponding to a MW of 382, 586, and 590, were not found in previous works in which the diamondoids were isolated from petroleum (Dahl et al., 2003a). These molecules were detected at high GC temperature, higher than the maximum GC temperature used by another group (593 K). For example, molecular ion peaks corresponding to a MW of 586 appeared when the GC oven temperature was raised to 620 K. In GC-MS measurements, each diamondoid is expected to elute at a distinct retention time. In contrast, for higher diamondoids, there are MW groups that have many structural isomers. For example, MW 292 (tetramantane) has 4 possible isomers and MW 396

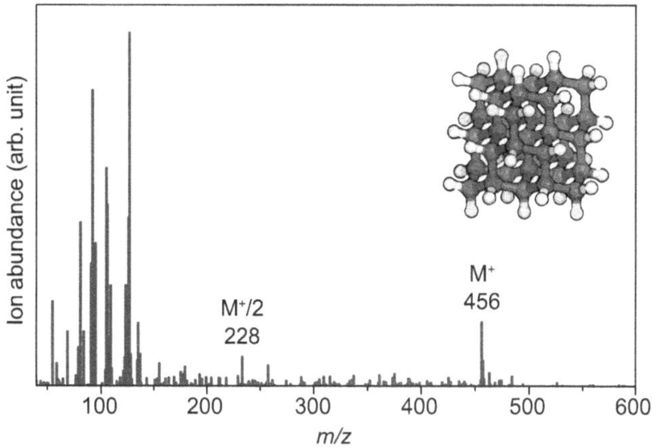

Figure 8.8 GC-MS mass spectrum of possible superadamantane (MW 456) obtained by PLA in $scCO_2$. The inset shows the molecular structure of superadamantane. Data adapted from Nakahara et al. (2011).

(hexamantane) has 28. Therefore, in the GC-MS measurements, diamond molecules corresponding to the same MW group but possessing different structures are expected to elute at different retention times. However, each MW group of diamondoid had only one or a few points per analysis in this study, even for those with a large number of possible structural isomers. Considering the relatively fast heating rate of GC-MS (20 $K.min^{-1}$) and the slow scan rate at 1.79 $scans.s^{-1}$ compared to those of other groups (10 $K.min^{-1}$ and 2.94 $scans.s^{-1}$) (Dahl et al., 2003a), the resolution of the GC-MS analyses was about 30% of the other groups' analyses. Therefore, it is expected that different isomers could appear at the same scan times, even if they elute at different times. Slower heating rates or faster scan rates will be used in future studies to improve the resolution of GC-MS analyses. Figure 8.8 shows an example of mass spectra corresponding to a MW of 456. The diamondoid corresponding to MW 456 is the most compact decamantane ($n = 10$), [1231241(2)3] decamantane, also known as superadamantane.

The spectrum shows a typical feature of unsubstituted higher diamondoids, that is, a strong molecular ion peak M^+ at m/z 456 with a few fragment ions including the $M^{+\cdot}_{/2}$ peak at $m/z \approx 228$, which corresponds to the doubly charged molecular ion. Strong peaks at low m/z in Fig. 8.8 are supposed to be due either to other hydrocarbon molecules such as paraffin and aromatics that co-elute at the same retention time, or to other materials that originated from the GC column and the septum that appear in the entire GC-MS analysis. The results of the GC-MS analyses can be summarized as follows: (i) There were many possible diamondoids synthesized, and their GC-MS retention times generally increase with the number of cages; (ii) some of the synthesized diamondoids have not been reported before, which might be due to the high GC oven temperature of up to 633 K; (iii) all MW groups of diamondoids had only one or a few points per analysis, which could be explained by the lower scanning resolution of the GC-MS analyses; and (iv) the mass spectra showed features that were similar to those of previously reported diamondoids with high MW, that is, a large molecular ion peak and a small number of fragment peaks.

8.4.3 Effects of Pyrolysis on Synthesized Products

The mass spectra acquired after the pyrolysis experiments and corresponding to possible higher diamondoids are listed in Tables 8.5 and 8.6.

The observed peaks were different before and after pyrolysis. However, the molecular ion peak with a MW of 456 was observed both before and after pyrolysis. Moreover, the relative retention

Table 8.5 List of observed higher diamondoids for PLA in $scCO_2$ after pyrolysis

Cage number (n)	Molecular formula	M^+ (m/z) base peak	GC-MS relative retention time
5	$C_{25}H_{30}$	330	1.70–2.25
5	$C_{26}H_{32}$	344	2.30–2.31
7	$C_{30}H_{34}$	394	2.02
7	$C_{34}H_{40}$	448	2.45
9	$C_{34}H_{36}$	444	1.66
9	$C_{37}H_{40}$	484	2.20
10	$C_{35}H_{36}$	456	2.12–2.13
10	$C_{44}H_{48}$	576	2.57

Table 8.6 List of observed higher diamondoids for PLA in $scCO_2$ with cyclohexane after pyrolysis

Cage number (n)	Molecular formula	M^+ (m/z) base peak	GC-MS relative retention time
4	$C_{22}H_{28}$	292	1.34
5	$C_{26}H_{32}$	344	1.83
6	$C_{30}H_{36}$	396	1.67
7	$C_{32}H_{36}$	420	2.01
7	$C_{34}H_{40}$	448	2.45

Data adapted from Nakahara et al. (2011).

times of GC-MS were very close to each other. This result suggests that the synthesized products with a MW of 456 survived after pyrolysis. It is expected that nondiamondoid hydrocarbons of high MW will decompose after thermal treatment at 723 K. Therefore, the present results suggest that the peak corresponding to a MW of 456 originated from superadamantane. Many MW groups of diamondoids such as those with $n = 5$ and 7 cages, were not found before pyrolysis, although they were found after the pyrolysis experiments.

In $scCO_2$, species originating from the decomposition of CO_2 could possibly participate in the formation of new materials. Assuming those materials were nondiamondoid hydrocarbons and were decomposed after pyrolysis, this might explain why the observed spectra contained peaks corresponding to diamondoids with five and seven cages only after pyrolysis.

8.5 Comparison between PLA in $scCO_2$ and scXe

There are MW groups that were only synthesized in scXe and not in $scCO_2$ (Nakahara et al., 2010). Moreover, the production rate of diamantane of PLA in $scCO_2$ was 0.2 $\mu g h^{-1}$, which is lower than that of PLA in scXe. This could be attributed to the lower solubility of adamantane in $scCO_2$ compared with that of scXe. The production rate of diamantane is approximately the same in neat $scCO_2$ and in the $scCO_2$/cyclohexane mixture. Therefore, while the solubility of adamantane in $scCO_2$ can be enhanced using cyclohexane as a co-solvent, this is probably not the only factor for increasing the production rate. In the case of scXe, the molecules of the medium do not participate in the reaction, whereas in the case of $scCO_2$, CO_2 itself can also dissociate and form reaction products. Consequently it is suspected that the lower reaction yield of diamantane in $scCO_2$ is due to both the lower solubility of adamantane and competing reactions resulting in the formation of nondiamondoids. In contrast, the mass spectrum showing a MW of 456 was only obtained by PLA in pure $scCO_2$. C—C and C—H bonds must be dissociated to allow the formation of successive diamondoids from adamantane. In the previous work where diamondoids were synthesized by DBD in scXe (Stauss et al., 2010), it was suggested that the absence of oxidants resulted in a preferential dissociation of C—C bonds and led to the synthesis of diamondoids with a high H/C ratio. On the contrary, the products from PLA in $scCO_2$ include relatively low H/C ratio diamondoids compared with those in scXe. It is supposed that oxidant species originating from $scCO_2$ might lead to selective dissociation of C—H bonds, enabling the synthesis of low H/C ratio diamondoids, such as superadamantane.

8.6 Conclusions and Perspectives

In this chapter, we have reviewed diamondoids; their structure, physical and chemical properties; and a short overview of their current main and possible future applications. While smaller diamondoids up to tetramantane can be synthesized by conventional organic chemical methods, the synthesis beyond tetramantane has turned out to be not possible.

PLA was performed in $scCO_2$ with and without cyclohexane as a cosolvent. Raman spectra of the synthesized products indicated the presence of sp^3-hybridized materials including diamondoids. From GC-MS measurements, this indicates the synthesis of diamantane and possibly other, higher-order diamondoids, including those with number of cages larger than reported so far ($n = 12$) and superadamantane ($n = 10$). Because oxidant species originating from $scCO_2$ might lead to the selective dissociation of C—H bonds, synthesis of higher-order diamondoids, such as superadamantane, could be realized. Moreover, the authors found that there were more higher-order diamondoids with $n = 5$–10 for PLA in $scCO_2$. In summary, PLA in SCFs and in particular $scCO_2$ is considered to be a promising approach for the synthesis of higher-order diamondoids.

Chapter 9

Synthesis of Diamondoids by Atmospheric-Pressure Microplasmas

This chapter addresses a more recent alternative approach for the synthesis of diamondoids. As discussed in Chapters 7 and 8, both electric discharge and pulsed laser plasmas generated in supercritical fluids (SCFs) allow the synthesis of diamondoids. However, one disadvantage of the use of these types of plasmas is the necessity to use specialized high-pressure equipment. In this chapter, we introduce a recent alternative approach, which is based on the use of microplasma reactors that are operated at atmospheric pressure. After a brief review of atmospheric-pressure microplasma sources, we give a short overview of the results.

9.1 Introduction

As has been shown in the previous chapters, the synthesis of diamondoids is not straightforward, and the quantities that can be realized are not sufficient for application in industry.

Specifically, we have shown that nanodiamonds (Kikuchi et al., 2010) and diamondoids can be synthesized both by electric

Diamondoids: Synthesis, Properties, and Applications
Sven Strauss and Kazuo Terashima

ISBN 978-981-4745-18-5 (Hardcover), 978-1-315-36466-7 (eBook)
www.panstanford.com

discharges (Shizuno et al., 2011; Stauss et al., 2010) and by pulsed laser ablation plasmas in supercritical xenon (Nakahara et al., 2010) and CO_2 (Nakahara et al., 2011).

These previous studies were conducted using a batch-type process, that is, the reactions are realized keeping the system closed during the synthesis, without any in- or outflow. However, it has been shown that often continuous-flow microreactors, where the precursors are continuously introduced into the reactor, can yield better reproducibility and higher reaction yields (Hartman and Jensen, 2009). The reason is that in microfluidic reactors, thermal control is easier compared to macroscale reactor systems. The first combination of a microplasma and a high-pressure microfluidic reactor was demonstrated recently (Oshima et al., 2012, 2014), the system allowing increasing the reaction yields of diamondoids compared to those of previously used macroreactors. However, the realization of such high-pressure microplasma reactors is not straightforward, which is mainly due to the specialized high-pressure equipment and the high-pressure fittings.

On the other hand, the possibility of rapid, continuous gas-phase growth of semiconductor nanowires has been demonstrated (Heurlin et al., 2012), and the growth speed and morphology were shown to be influenced by the flow rate and precursor size. In addition, continuous-flow microplasma reactors operated at atmospheric pressure have been used for the synthesis of metallic nanoparticles (Chiang et al., 2010; Lin et al., 2012).

As an alternative to high-pressure microfluidic microplasmas, an atmospheric-pressure continuous-flow process using dielectric barrier discharge (DBD) plasma microreactors for the synthesis of diamondoids was developed. In this system, again adamantane was used as a precursor and seed, employing gases used in conventional chemical vapor deposition (CVD) diamond growth (Bachmann et al., 1991), that is, methane (CH_3) and hydrogen (H_2). In the following sections, we will first give a brief overview of atmospheric-pressure microplasma reactors that were used for the synthesis of metallic nanoparticles and nanodiamonds, before introducing the microreactor system used for the synthesis of diamondoids.

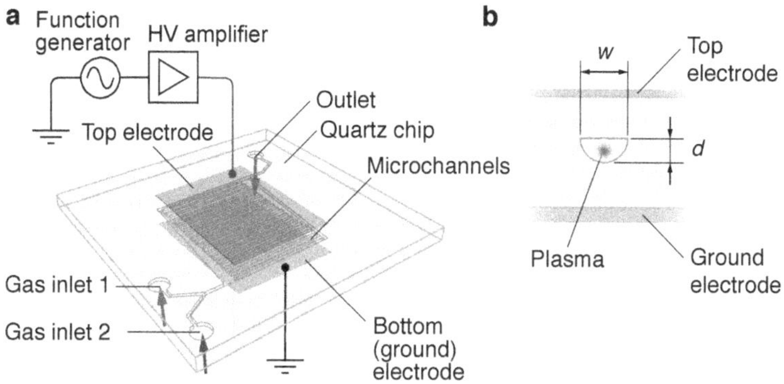

Figure 9.1 Geometry of a microchip microplasma reactor and its cross section. (a) Three-dimensional view of a quartz microreactor containing 20 microchannels. Gas inlets 1 and 2 were used for introducing the different types of source gases, including the precursor adamantane, into the microplasma reactor. (b) Cross section of a single micoreactor channel with indium tin oxide (ITO) as the top electrode, and a grounded bottom electrode. The channel widths *w* and depths *d* of the microplasma reactors ranged from 400 to 500 μm, and 100 to 200 μm, respectively. Reprinted from Ishii, C., Stauss, S., Kuribara, K., Urabe, K., Sasaki, T., and Terashima, K. (2015). Atmospheric-pressure synthesis of diamondoids by plasmas generated inside a microfluidic reactor, *Diamond and Related Materials* **59**, pp. 40–46, Copyright (2015), with permission from Elsevier.

9.2 Microchip Microplasma Reactors

Figure 9.1 displays the schematic experimental setup that was employed for diamondoid synthesis using microplasma reactors. Here, custom-made chip microreactors fabricated out of glass were used that were realized by using lithography and powder blasting. The top and bottom electrodes consisted of an indium tin oxide film (to have optical access to the reactor), while the bottom electrode consisted of silver.

To realize the glass microreactors, self-adhesive UV photosensitive films and inkjet-printed transparency masks were used. This allowed achieving fluidic channels with widths of 400 $\pm$ 50 μm and depths of 250 $\pm$ 50 μm).

For the synthesis experiments, the microchip reactors were fixed inside a custom-made compression holder similar to those developed for high-pressure microfluidic reactors (Marre et al., 2010).

The temperature of the cylinder that contained the precursor adamantane, the tubing system, and the microreactor was adjusted using resistive heaters. Prior to each synthesis experiment and before flowing the reactant gases into the cell, the system was evacuated using a rotary pump. The temperature of the pipes and the reactor and the pressure at the inlet of the reactor were monitored using temperature and pressure indicators. The electric discharges were generated by applying a sine wave signal produced by a function generator and an alternating current (AC) high-voltage amplifier.

The synthesis experiments were conducted for durations of 120 or 240 min at a frequency of $f = 10$ kHz, and applied peak-to-peak voltages that varied between 3.2–4.3 $kV_{p\text{-}p}$ (cf. Table 9.1 for the detailed experimental conditions). To investigate the influence of the gas composition, in particular the addition of hydrogen (H_2) and methane (CH_4), on the synthesis of diamondoids, we used four different gas mixtures: adamantane (Ada) vapor/1% CH_4/1% H_2/argon (Ar), Ada/0.5% CH_4/Ar, Ada/1% H_2/Ar, and Ada/Ar.

The gas and the synthesized products were first flown through a collection trap filled with cyclohexane, and then a solid-phase extraction filter, to collect products that might not have been retained by the cyclohexane trap.

After the synthesis, the tubes and the microreactor were flushed by pumping cyclohexane through the whole system. The solid-phase extraction filters were also flushed separately by pumping cyclohexane through them and the solutions collected. To increase the concentration of the products for analysis, the cyclohexane was first evaporated on a hot plate at a temperature of 85°C before adding 1 mL of cyclohexane to redissolve the products.

The collected samples were then characterized by gas chromatography–mass spectrometry (GC-MS). In addition to the total ion chromatograms (TICs), selected ion monitoring (SIM) chromatograms corresponding to molecular masses of diamondoids in the mass-to-charge (m/z) range of 136–600, and those of possible reaction intermediates up to m/z 300 were acquired. Individual mass spectra were also measured in the scan mode over an

Table 9.1 Experimental conditions adopted for diamondoid synthesis (Ishii et al., 2015). The concentration of the precursor adamantane ρ is calculated from the expression of the sublimation pressure of adamantane by Bazyleva et al. (2011), under the assumption of a perfect gas. For a temperature of 300 K, the adamantane sublimation pressure is 0.02 kPa, and at 320 K, it is about 0.09 kPa (i.e. for both temperatures, the adamantane content in the gas mixtures was below 0.1%). The residence time of about 0.1 s was estimated from the fixed flow rate (20 sccm) and the reactor geometry

Parameters	Symbol	Values	
Argon	Ar	92–100	%
Methane	CH_4	0–4	%
Hydrogen	H_2	0–4	%
Total flow rate	Q_{tot}	20	sccm
Residence time	t_{res}	~0.1	s
Average temperature	T	300, 320	K
Pressure	p	100	kPa
Applied voltage	V_{appl}	3.2 – 4.3	$kV_{p\text{-}p}$
Applied frequency	f	10	kHz
Adamantane concentration	ρ	~1.2, ~4.7	$g\,m^{-3}$
Plasma generation time	t_p	120, 240	min

m/z range of 40–1000. For the identification of possible reaction intermediates, we compared the retention times of peaks in the TIC and SIM data to those of diamondoids and derivates isolated from crude oil (Liang et al., 2012) and by matching of the experimental mass spectra to reference mass spectra (NIST/EPA/NIH Mass Spectral Library (EI), 2012).

In addition to the analysis of the synthesized products by GC-MS, the authors also monitored the plasmas by current–voltage (I–V) and charge-voltage (Q–V) measurements and optical emission spectroscopy (OES).

9.3 Plasma Generation and Characterization

Figure 9.2 shows an example of a microreactor containing 20 parallel microchannels with photographs prior to (Fig. 9.2a) and during

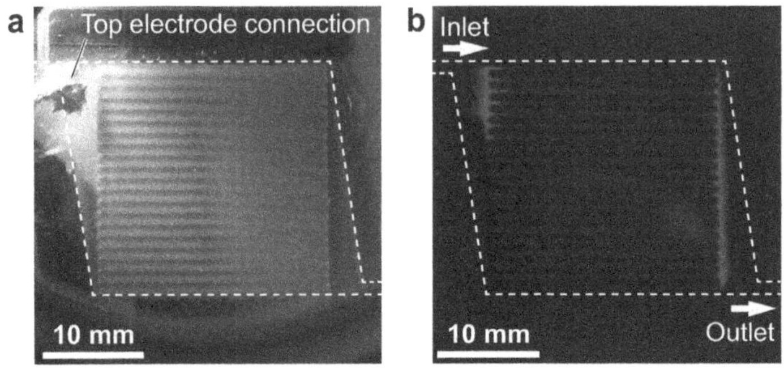

Figure 9.2 Photographs of microplasma reactor during synthesis. (a) Top view through ITO electrode of reactor prior to plasma generation. The dotted lines indicate the geometries of the microchannels and the positions of the inlet and outlet. (b) Photograph of reactor during plasma generation (experimental conditions: $T = 320\,\mathrm{K}$; $V_{\mathrm{appl}} = 3.6–3.7\,\mathrm{kV_{p\text{-}p}}$; $f = 10\,\mathrm{kHz}$). Reprinted from Ishii, C., Stauss, S., Kuribara, K., Urabe, K., Sasaki, T., and Terashima, K. (2015). Atmospheric-pressure synthesis of diamondoids by plasmas generated inside a microfluidic reactor, *Diamond and Related Materials* **59**, pp. 40–46, Copyright (2015), with permission from Elsevier.

(Fig. 9.2b) plasma generation. The trapezoidal geometry of the inlet and outlet zones, as indicated in the photograph in Fig. 9.2a, allow one to achieve approximately the same gas velocity distribution in all microchannels (Commenge et al., 2002).

The discharge current (I_{d}) was calculated from the measured current (I_{m}) and applied voltage (V_{appl}) curves (Stauss et al., 2013). Typical capacitances for estimating I_{d} were of the order of $C_{\mathrm{g}} = 6 \pm 2\,\mathrm{pF}$ (capacitance due to gas gap), $C_{\mathrm{sd}} = 14 \pm 3\,\mathrm{pF}$ (capacitance of glass slides), and $C_{\mathrm{p}} = 11 \pm 2\,\mathrm{pF}$ (parasitic capacitance). In the current–voltage waveforms, a broad current peak in the positive and negative half-cycles appears, the (absolute) maximum current being of the order of $\sim$2 mA. In this case, because of the very low current, individual discharge peaks inside this broad discharge peak could not be observed.

The power consumption for this discharge was found to be $P = 134 \pm 1\,\mathrm{mW.cm^{-2}}$.

In general, for the applied voltages used for the synthesis, the average power consumption of the discharges was found to be in the range of $\sim$130–200 $\mathrm{mW.cm^{-2}}$. The variations are mainly attributed to the different gas compositions and differences in the channel depths.

9.3.1 Optical Emission Spectroscopy Measurements

Figure 9.3 shows optical emission spectra of the discharges generated for four experimental conditions where the levels of H_2 and CH_4 were set to values between 0% and 1%, respectively.

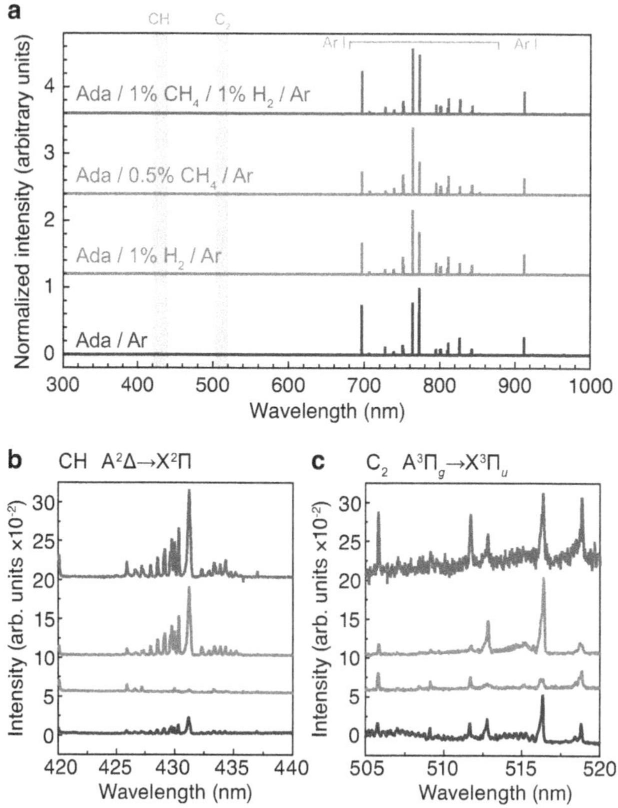

Figure 9.3 Optical emission spectra of four different experimental conditions. (a) Overview in the wavelength range of 300–1000 nm. (b) CH $A^2\Delta \rightarrow X^2\Pi$ band at 431.4 nm. (c) C_2 $A^3\Pi_g \rightarrow X^3\Pi_u$ Swan band at 516.5 nm. The detailed spectra in (b) and (c) were normalized with respect to the 811 nm Ar I line. Data adapted from Stauss et al. (2014).

The topmost spectrum (Fig. 9.3a) shows an overview of the plasma emission in the wavelength range between 300 and 1000 nm for the four experimental conditions indicated in the graph. The most prominent peaks are due to argon neutral radicals (Ar I) in the range of 680–850 nm.

To investigate the effect of the gas composition on the optical emission spectra, CH and C_2 bands and the atomic hydrogen lines at 486.1 (H_β) and 656.3 nm (H_α) were monitored. Figure 9.3a shows a detailed view of the CH $A^2\Delta \rightarrow X^2\Pi$ band at 431.4 nm and Fig. 9.3b of the C_2 $A^3\Pi_g \rightarrow X^3\Pi_u$ Swan band at 516.5 nm.

In the case of the Ada/Ar mixture, atomic H emission could not be detected, whereas in all the other compositions, only the H_α line could be observed, but its intensity was very weak. In a previous study, OES measurements during the synthesis of diamond films using direct current microplasma discharges at pressures between 100 and 500 torr (Sankaran and Giapis, 2002) did not show the presence of C_2 bands. However, the authors observed strong H_α and H_β emission. Here, as shown in Fig. 9.3b,c, when CH_4 is added to the gas, both CH and C_2 bands were observed, which are attributed to the dissociation of adamantane and CH_4. In the case of the Ada/1% H_2/Ar mixture, the intensities of these two bands are very weak, whereas for Ada/Ar, the intensity is higher. The reason could be that in the absence of H_2, almost all of the plasma's energy is used primarily for the dissociation of adamantane.

It has been shown previously that in the growth of ultrananocrystalline diamond, C_2 emission is stronger, whereas for the synthesis of nanocrystalline diamond, CH bands are dominant. As will be discussed later, the presence of CH and C_2 can probably attributed to both the dissociation of adamantane and CH_4. Therefore, for achieving higher reaction yields, it might be advantageous to realize plasma conditions where the adamantane cage is preserved and only precursor gases that act as a carbon source are dissociated.

In addition, in conventional CVD growth of diamond, the presence of atomic hydrogen is crucial (May et al., 2006). Therefore, it is expected that higher H_2 concentrations than those employed in the studies of Stauss et al. (2014) and Ishii et al. (2015) could enable more efficient diamondoid synthesis.

9.3.2 GC-MS Analysis of Diamantane and Reaction Intermediates

Since OES does not allow quantitative analysis very easily, the prediction of reaction mechanisms is difficult, if not impossible. Furthermore, other species that are considered relevant for diamond synthesis, such as CH_3 and C_2H_2, are not observable in optical emission (Williams et al., 2006). Therefore, in addition to OES measurements and to understand the influence of the gas composition on diamondoid synthesis, we have also investigated the presence of possible reaction intermediates by GC-MS measurements.

Figure 9.4 shows mass spectra of the synthesized product and a standard sample that contained a mixture of adamantane and the second diamondoid, diamantane, both acquired at the same retention time $t_{ret} = 8.3$ min. The sample (Fig. 9.4a) was synthesized at conditions of $T = 300$ K, $V_{appl} = 4.0\,kV_{p\text{-}p}$, and $f = 10$ kHz, in a mixture of Ada/4% CH_4/Ar.

The mass spectrum of the synthesized product and the one of the standard sample (Fig. 9.4b) both contain a prominent peak at m/z 188, which is the molecular ion peak of diamantane, which has a relative molecular mass (M_r) of 188. In addition to the acquisition of the mass spectrum at the same retention time and the presence of the molecular ion peak at m/z 188, both mass spectra have similar fragmentation patterns (the smaller differences are attributed to the lower quantity of the synthesized sample when compared to the standard sample). This shows that diamantane can be synthesized from adamantane by atmospheric-pressure microplasmas.

Figure 9.5 displays the TICs of products collected in the cyclohexane trap and obtained using four different gas mixtures, for retention times between 4.1 and 7.8 min, that is, before the elution of diamantane ($t_{ret} = 8.3$ min).

The large broad peak at $t_{ret} \approx 3.9$–4.1 min is due to the precursor adamantane. At a retention time of about 3.3–3.4 min in all gas mixtures, one can observe a prominent peak that is ascribed to 1-methyladamantane (1-MA). For all gas compositions, at $t_{ret} \approx 5.1$ min, the TICs also contain a peak corresponding to 2-methyladamantane (2-MA); however, this peak is most prominent for the cases including CH_4 (1% and 0.5%). Finally at $t_{ret} \approx 5.9$ min,

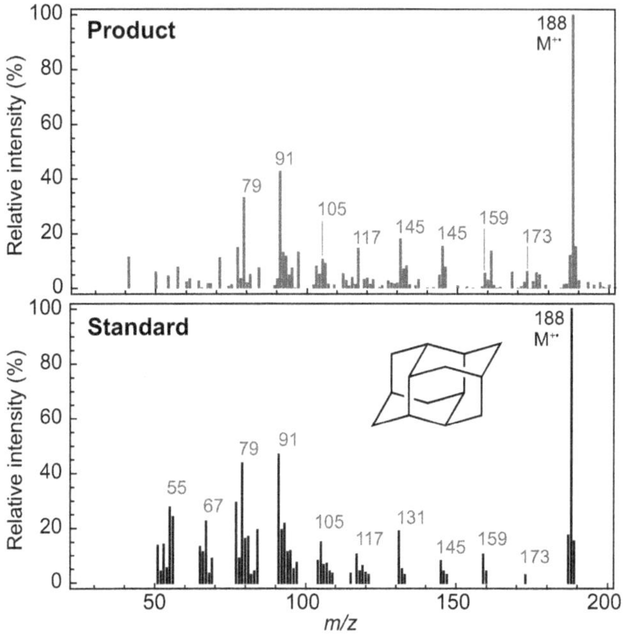

Figure 9.4 Mass spectra of synthesized product and standard diamantane sample. Both the product and reference sample were acquired at a retention time $t_{ret} = 8.3$ min for an adamantane/4% CH_4/Ar plasma. (a) The spectrum acquired of the synthesized product shows a base peak at m/z 188. (b) The standard sample shows a mass spectrum that has the same dominant peak at m/z 188, which is due to the molecular ion of diamantane ($M^{+\cdot}$). The fragment peaks and their distributions are also similar, which indicates the synthesis of diamantane. Synthesis conditions: $T = 300$ K (adamantane vapor pressure 0.02 kPa), $V_{appl} = 4.0\ kV_{p\text{-}p}$, $f = 10$ kHz, plasma generation time $t_p = 240$ min. Adapted from Stauss et al. (2014).

one can observe the appearance of a peak that can be attributed to 1,4-dimethyladamantane (1,4-DMA).

The integrated TIC peak intensities of the adamantane derivatives, 1-MA, 2-MA, and 1,4-DMA, acquired for the four different gas compositions are presented in Fig. 9.6. As can be seen in the graphs, the quantities of the investigated adamantane derivatives, which might play a crucial role as reaction intermediates in the synthesis of diamantane from adamantane in one possible proposed step-wise reaction mechanism (Oshima et al., 2014), highly depend on the

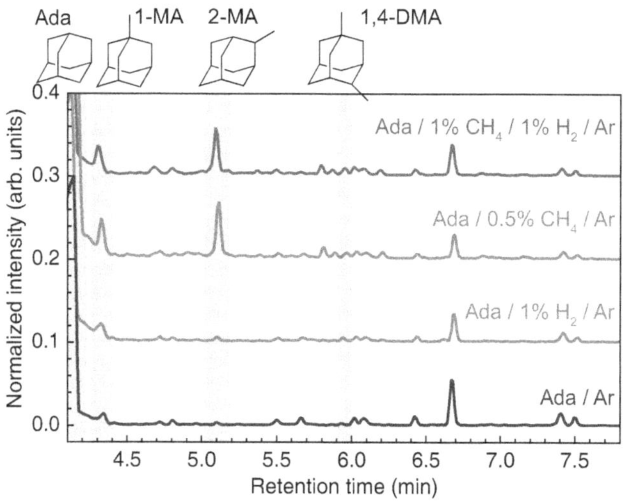

Figure 9.5 Total ion chromatograms of products synthesized for four different gas mixtures. The shaded regions in the graph indicate respectively the retention times of the precursor adamantane (Ada), 1-methyladamantane (1-MA), 2-methyladamantane (2-MA), and 1,4-dimethyladamantane (1,4-DMA). The corresponding molecular structures are also displayed on the top of the graph. Data adapted from Stauss et al. (2014).

gas chemistry. In the case when only Ar is present in the gas, the formation of reaction intermediates is very low. Consequently, for achieving higher reaction yields, it is necessary to adjust the plasma gas chemistry so that the two main intermediate reaction steps, that is, hydrogen abstraction and addition of alkyls, can proceed efficiently.

For the cases where no CH_4 or H_2 was added to the plasma gas, the intensities of methyladamantanes are close to zero. In addition, one can see that there is a competition in the formation of 1-MA and 2-MA. Considering the structure of diamantane and its formation starting from adamantane, 1-MA and similar derivatives, that is, those that form by addition of alkyls at tertiary C-positions in adamantane, are not desirable, because such molecular structures do not allow the formation of the next higher diamondoid (Oshima et al., 2014). However, the gas-phase enthalpy of formation is -170 ± 1 kJ.mol^{-1} for 1-MA and -154 kJ.mol^{-1} for 2-MA (Clark et al.,

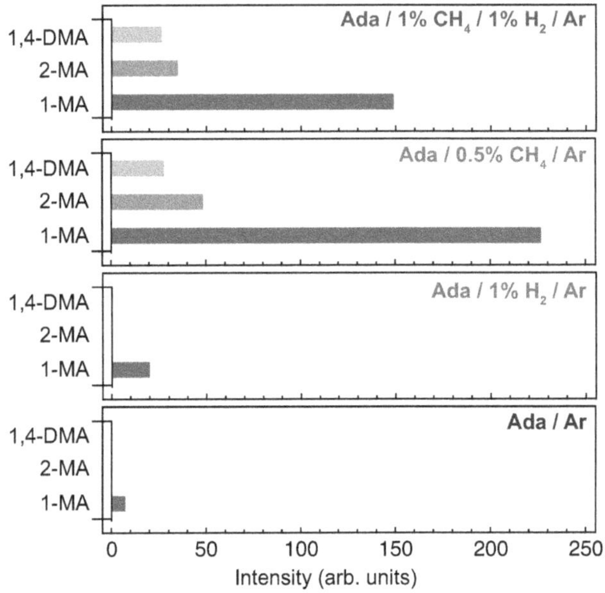

Figure 9.6 Intensities of adamantane derivatives 1-methyladamantane (1-MA), 2-methyladamantane (2-MA), and 1,4-dimethyladamantane (1,4-DMA) for four different plasma gas compositions. The intensity values were obtained by integration of the peak areas in the total ion chromatograms. Data adapted from Stauss et al. (2014).

1979), and from ab initio calculations the ground-state energy of the $1-\text{adamantyl}^+$ cation was found to be -10604.51 eV, which is more stable compared to $2-\text{adamantyl}^+$ (-10604.04 eV) (Steglich et al., 2011).

In addition, in the case of the Ada/Ar mixture, the presence of CH bands in the OES and the appearance of adamantane derivatives in the GC-MS measurements indicate that the precursor adamantane is dissociated by the discharge. The bonding energy of the (sp^3)C–C(sp^3) bonds in adamantane is weaker than the (sp^3)C–H bonds in both adamantane (Shen et al., 1992) and CH_4, and consequently, the dissociation of the C–C bonds by the plasma is more probable in comparison to cleaving of the C–H bonds.

Therefore it is surmised that even when 2-MA is formed, it might not persist in sufficient quantities to allow further hydrogen abstraction and subsequent carbon radical addition. Consequently, assuming that one possible reaction mechanism of diamantane growth from adamantane at atmospheric pressure consists of repeated hydrogen abstraction and methyl attachment (Oshima et al., 2014), conditions that promote the formation and persistence of reaction intermediates, that is, 2-MA, 2,4-DMA, and 2,4,9-trimethyladamantane have to be established.

The quantities of diamantane obtained in the present study were of the order of a few nanograms, about 1 to 2 orders of magnitude smaller than what was obtained by synthesis in supercritical xenon, where the maximum quantity of diamantane obtained was on the order of micrograms (Oshima et al., 2014). One reason for the lower yields could be the lower concentration of the precursor adamantane at atmospheric pressure: For the temperatures employed in the present study, the estimated adamantane concentrations were $\sim$1.2 and 4.7 g.m^{-3}, whereas for the plasma microreactor using supercritical xenon near its critical point allowed to dissolve adamantane of the order of $\sim$2000 g.m^{-3}. Furthermore, it is surmised that in the case of plasmas generated in supercritical conditions, the presence of high-density clusters containing ions or neutral radicals aids in the reaction of diamondoids.

9.4 Summary

In this chapter, we have presented alternative approaches for the synthesis of diamondoids, which are based on plasmas generated in different gases with pressures ranging from atmospheric conditions up to SCFs.

DBDs allowed us to synthesize diamantane using adamantane as a precursor and seed. Similar to what has been observed in liquid-based chemistry synthesis, it seems that when compared to conventional batch-type reactors, the efficiency of reactions can be enhanced by using microscale reactors. Investigation of reaction mechanisms by GC-MS indicate that one possible reaction mechanism consists in repeated hydrogen abstraction and addition of methyl groups.

We have also presented research on plasmas generated in SCFs and their application to the synthesis of diamondoids. While much easier to realize because there is no need to control the electrode geometry, currently there are no easy means to scale up pulsed laser ablation processes that would allow increasing the quantities of diamondoids synthesized.

Finally, we have demonstrated the possibility of employing DBD microfluidic reactors for the synthesis of diamondoids at atmospheric pressure. One possible reaction mechanism of larger diamondoids from adamantane is considered to take place in a stepwise reaction that consists of repeated hydrogen abstraction and addition of carbon-containing fragments such as alkyls. GC-MS measurements showed that the formation of possible intermediates highly depends on the gas composition, namely the presence of species that can provide carbon to form next larger diamondoids.

The present results show that diamondoids can be synthesized at atmospheric pressure, albeit in lower quantities than in previously employed plasmas generated in high-pressure and supercritical conditions. However, it is suggested that by adjusting the plasma chemistry, higher reaction yields could be realized. It is also surmised that by employing much higher frequencies (i.e., in the ultrahigh frequency or microwave range) to generate the plasmas, the quantities of synthesized products could be increased.

Compared to conventional organic chemistry synthesis, while very promising, the approaches presented in this chapter are still in their infancy. For further improving diamondoid synthesis by these different, plasma-based processes, more in-depth characterization of the products is necessary, not only after collection, but also during synthesis. Therefore, online measurement of the plasma species by MS might help to further advance the proposed synthesis process.

Chapter 10

Conclusions and Perspectives

With this chapter, we conclude our journey into the field of diamondoids. We started it by presenting the reader with an introduction about diamondoids, their molecular structure, and their naming conventions and also by introducing their main chemical and physical properties, including their electronic structure and optical and vibrational properties. We have also given a short overview of their current and possible future applications, which span a wide range of different scientific and technological fields: medicine, pharmaceutics, biotechnology, materials science, and opto- and nanoelectronics.

In addition, we have briefly talked about how diamondoids can form in natural gas and oil reservoirs and the methods to isolate and purify them. We have also discussed the currently known approaches to obtaining diamondoids by conventional organic chemical synthesis.

At the moment, only the two smallest diamondoids, adamantane and diamantane, are available in quantities large enough (tons/year) so that they can be used on an industrial scale. While higher diamondoids up to undecamantane have been found in oil reservoirs, their availability is limited to research purposes only.

Diamondoids: Synthesis, Properties, and Applications
Sven Strauss and Kazuo Terashima

ISBN 978-981-4745-18-5 (Hardcover), 978-1-315-36466-7 (eBook)
www.panstanford.com

On the basis of the current knowledge of the possibilities of organic chemistry and the unsuccessful attempts to obtain diamondoids beyond triamantane by these methods, it seems that alternative synthesis approaches are desperately needed in order to realize the fabrication of higher diamondoids in quantities useful for practical applications.

As we have shown in the last part of this book, possibilities exist that allow synthesizing of diamondoids both by electric discharge and by pulsed laser plasmas generated in supercritical fluids. Moreover, in principle, diamondoids can also be produced by plasmas generated in atmospheric-pressure plasma microreactors. However, compared to the synthesis in supercritical fluids, the quantities obtained by this approach are much lower.

To advance the field of diamondoids and extend their use in various applications, it is necessary to further investigate possibilities for their synthesis. We hope that the present text can serve as an inspiration for both experienced and young researchers to engage in research on this fascinating class of nanomaterials so that the field of diamondoids and also their applications can grow.

Appendix

A.1 Character Code Tables of Diamondoids

For the sake of completeness, we list the character tables of the different point groups of the most common diamondoids.

Table A.1 Character code table for the C_1 point group

C_1	E	$h = 1$
A	1	

Table A.2 Character code table for the C_s point group

C_s	E	σ_h	$h = 2$	
A′	1	1	x, y, R_z	x^2, y^2, z^2, xy
A″	1	−1	z, R_x, R_y	yz, zx

Table A.3 Character code table for the C_2 point group

C_2	E	C_2	$h = 2$	
A	1	1	z, R_z	x^2, y^2, z^2, xy
B	1	−1	x, y, R_x, R_y	yz, zx

Table A.4 Character code table for the C_{2v} point group

C_{2v}	E	C_2	$\sigma_v(xz)$	$\sigma_v'(yz)$	$h = 4$	
A_1	1	1	1	1	z	x^2, y^2, z^2
A_2	1	1	−1	−1	R_z	xy
B_1	1	−1	1	−1	x, R_y	zx
B_2	1	−1	−1	1	y, R_x	yz

Table A.5 Character code table for the C_{2h} point group

C_{2h}	E	$C_2(z)$	i	σ_h		$h = 4$
A_g	1	1	1	1	R_z	x^2, y^2, z^2, xy
B_g	1	−1	1	−1	R_x, R_y	xz, yz
A_u	1	1	−1	−1	z	
B_u	1	−1	−1	1	x, y	

Table A.6 Character code table for the D_{3d} point group

D_{3d}	E	$2C_3$	$3C_2$	i	$2S_6$	$3\sigma_d$		$h = 12$
A_{1g}	1	1	1	1	1	1		$x^2 + y^2, z^2$
A_{2g}	1	1	−1	1	1	−1	R_z	
E_g	2	−1	0	2	−1	0	(R_x, R_y)	$(x^2 - y^2, xy)$ (zx, yz)
A_{1u}	1	1	1	−1	−1	−1		
A_{2u}	1	1	−1	−1	−1	1	z	
E_u	2	−1	0	−2	1	0	(x, y)	

Table A.7 Character code table for the T_d point group

D_{3d}	E	$8C_3$	$3C_2$	$6S_4$	$6\sigma_d$		$h = 24$
A_1	1	1	1	1	1		$x^2 + y^2 + z^2$
A_2	1	1	1	−1	−1		
E	2	−1	2	0	0		$(2z^2 - x^2 - y^2, x^2 - y^2)$
T_1	3	0	−1	1	−1	(R_x, R_y, R_z)	
T_2	3	0	−1	−1	1	(x, y, z)	(xy, yz, zx)

Bibliography

Aczel, T., Gorbaty, M. L., Maa, P. S., and Schlosberg, R. H. (1979). Stability of adamantane and its derivatives to coal-liquefaction conditions, and its implications toward the organic structure of coal, *Fuel* **58**, 3, pp. 228–230.

Aharonovich, I., Greentree, A. D., and Prawer, S. (2011). Diamond photonics, *Nature Photonics* **5**, 7, pp. 397–405.

Amans, D., Chenus, A. C., Ledoux, G., Dujardin, C., Reynaud, C., Sublemontier, O., Masenelli-Varlot, K., and Guillois, O. (2009). Nanodiamond synthesis by pulsed laser ablation in liquids, *Diamond and Related Materials* **18**, 2–3, pp. 177–180.

Ambrose, D., Ewing, M., and McGlashan, M. (2014). Critical Constants and Second Virial Coefficients of Gases, http://www.kayelaby.npl.co.uk, http://www.kayelaby.npl.co.uk/chemistry/3_5/3_5.html.

Amendola, V., and Meneghetti, M. (2009). Laser ablation synthesis in solution and size manipulation of noble metal nanoparticles, *Phys. Chem. Chem. Phys.* **11**, 20, pp. 3805–3821.

Anastas, P. T., and Kirchhoff, M. M. (2002). Origins, current status, and future challenges of green chemistry, *Accounts of Chemical Research* **35**, 9, pp. 686–694.

Arai, Y., Sako, T., and Takebayashi, Y. (eds.) (2001). *Supercritical Fluids: Molecular Interactions, Physical Properties and New Applications*, Materials Processing (Springer).

Ashfold, M. N. R., Claeyssens, F., Fuge, G. M., and Henley, S. J. (2004). Pulsed laser ablation and deposition of thin films, *Chemical Society Reviews* **33**, 1, pp. 23–31.

Atkins, P., Overton, T., Rourke, J., Weller, M., and Armstrong, F. (2010). *Shriver and Atkin's Inorganic Chemistry*, 5th ed. (W.H. Freeman, New York).

Bachmann, P. K., Leers, D., and Lydtin, H. (1991). Towards a general concept of diamond chemical vapor-deposition, *Diamond and Related Materials* **1**, 1, pp. 1–12.

Badziag, P., Verwoerd, W. S., Ellis, W. P., and Greiner, N. R. (1990). Nanometre-sized diamonds are more stable than graphite, *Nature* **343**, 6255, pp. 244–245.

Baeyer, A. (1900). Systematik und nomenclatur bicyclischer kohlenwasserstoffe, *Berichte der deutschen chemischen Gesellschaft* **33**, 3, pp. 3771–3775.

Bailey, R. T. (1971). Spectroscopic and thermodynamic studies of molecular crystals. 1. Internal vibrations of adamantane in plastic phase, *Spectrochimica Acta Part A: Molecular Spectroscopy* **A 27**, 8, pp. 1447–1453.

Balaban, A. T. (2012). Partitioned-formula periodic tables for diamond hydrocarbons (diamondoids), *Journal of Chemical Information and Modeling* **52**, 11, pp. 2856–2863.

Balaban, A. T., Klein, D. J., Dahl, J. E., and Carlson, R. M. K. (2007). Molecular descriptors for natural diamondoid hydrocarbons and quantitative structure-property relationships for their chromatographic data, *The Open Organic Chemistry Journal* **1**, 1, pp. 13–31.

Balaban, A. T., and Rucker, C. (2013). How to specify the structure of substituted blade-like zigzag diamondoids, *Central European Journal of Chemistry* **11**, 9, pp. 1423–1430.

Balaban, A. T., and Schleyer, P. V. (1978). Systematic classification and nomenclature of diamond hydrocarbons. 1. Graph-theoretical enumeration of polymantanes, *Tetrahedron* **34**, 24, pp. 3599–3609.

Ballutaud, D., Jomard, F., Kociniewski, T., Rzepka, E., Girard, H., and Saada, S. (2008). Sp(3)/sp(2) character of the carbon and hydrogen configuration in micro- and nanocrystalline diamond, *Diamond and Related Materials* **17**, 4–5, pp. 451–456.

Banerjee, S., and Saalfrank, P. (2014). Vibrationally resolved absorption, emission and resonance raman spectra of diamondoids: a study based on time-dependent correlation functions, *Physical Chemistry Chemical Physics* **16**, 1, pp. 144–158.

Bazyleva, A. B., Blokhin, A. V., Kabo, G. J., Charapennikau, M. B., Emel'yanenko, V. N., Verevkin, S. P., and Diky, V. (2011). Thermodynamic properties of adamantane revisited, *Journal of Physical Chemistry B* **115**, 33, pp. 10064–10072.

Becker, K. H., Kogelschatz, U., Schoenbach, K. H., and Barker, R. J. (2004). *Non-equilibrium Air Plasmas at Atmospheric Pressure* (Taylor & Francis, London, United Kingdom).

Berwick, L., Alexander, R., and Pierce, K. (2011). Formation and reactions of alkyl adamantanes in sediments: carbon surface reactions, *Organic Geochemistry* **42**, 7, pp. 752–761.

Bistricic, L., Pejov, L., and Baranovic, G. (2002). A density functional theory analysis of Raman and IR spectra of 2-adamantanone, *Journal of Molecular Structure-Theochem* **594**, pp. 79–88.

Bjørlykke, K. (2010). *Petroleum Geoscience: From Sedimentary Environments to Rock Physics* (Springer, Berlin Heidelberg).

Blaney, F., Johnston, D. E., Mckervey, M. a., Jones, E. R. H., and Pragnell, J. (1974). Hydroxylation of diamantan-1-ol and diamantan-4-ol with fungus rhizopus-nigricans, *Journal of the Chemical Society, Chemical Communications*, 8, pp. 297–298.

Burns, W., McKervey, M. A., Mitchell, T. R. B., and Rooney, J. J. (1978). New approach to construction of diamondoid hydrocarbons: synthesis of anti-tetramantane, *Journal of the American Chemical Society* **100**, 3, pp. 906–911.

Burns, W., McKervey, M. A., and Rooney, J. J. (1975). New synthesis of triamantane involving a novel rearrangement of a polycyclic olefin in gas-phase on platinum, *Journal of the Chemical Society, Chemical Communications*, 24, pp. 965–966.

Burns, W., Mitchell, T. R. B., McKervey, M. A., Rooney, J. J., Ferguson, G., and Roberts, P. (1976). Gas-phase reactions on platinum: synthesis and crystal-structure of anti-tetramantane, a large diamondoid fragment, *Journal of the Chemical Society, Chemical Communications*, 21, pp. 893–895.

Cahill, P. A. (1990). Unusual degree of selectivity in diamantane derivatizations, *Tetrahedron Letters* **31**, 38, pp. 5417–5420.

Cernik, R., Evans, E. H. M., Hine, R., and Richards, J. P. G. (1978). Phase-transitions in triamantane, *Solid State Communications* **27**, 10, pp. 1017–1019.

Chakrabarti, J. K., Hotten, T. M., Sutton, S., and Tupper, D. E. (1976). Adamantane and protoadamantanealkanamines as potential anti-Parkinson agents. 10. *Journal of Medicinal Chemistry* **19**, 7, pp. 967–969.

Chan, Y. C., Choy, K. K. H., Chan, A. H. C., Ng, K. M., Liu, S. G., Sciamanna, S. F., Dahl, J. E., and Carlson, R. M. K. (2008). Solubility of diamantane, trimantane, tetramantane, and their derivatives in organic solvents, *Journal of Chemical and Engineering Data* **53**, 8, pp. 1767–1771.

Chang, Y. F., Chi, C. W., Chern, Y. T., and Wang, J. J. (2005). Effects of 1,6-bis[4-(4-amino-3-hydroxyphenoxy)phenyl]diamantane (dpd), a reactive oxygen species and apoptosis inducing agent, on human leukemia cells in vitro and in vivo, *Toxicology and Applied Pharmacology* **202**, 1, pp. 1–12.

Chen, Z. H., Tozaki, K., and Nishikawa, K. (1999). Development of thermal conductivity measurement for fluids which is convenient and effective for samples near the critical point, *Japanese Journal of Applied Physics* **38**, 1AB, pp. L92–L95.

Chern, Y. T. (1995). Synthesis and properties of new polycyclic polyesters from 1,6-diamantanedicarboxylic acyl chloride and aromatic diols, *Macromolecules* **28**, 16, pp. 5561–5566.

Chern, Y. T., and Huang, C. M. (1998a). Synthesis and characterization of new polyesters derived from 1,6- or 4,9-diamantanedicarboxylic acyl chlorides with aryl ether diols, *Polymer* **39**, 11, pp. 2325–2329.

Chern, Y. T., and Huang, C. M. (1998b). Synthesis and characterization of new polyimides derived from 4,9-diaminodiamantane, *Polymer* **39**, 25, pp. 6643–6648.

Chern, Y. T., and Wang, J. J. (1995a). Synthesis of 1,6-diamino-diamantane, *Tetrahedron Letters* **36**, 32, pp. 5805–5806.

Chern, Y. T., and Wang, W. L. (1995b). Synthesis and properties of new polyamides based on diamantane, *Macromolecules* **28**, 16, pp. 5554–5560.

Chern, Y. T., and Wang, W. L. (1996). Synthesis and characterization of tough polycyclic polyamides containing 4,9-diamantyl moieties in the main chain, *Journal of Polymer Science Part A: Polymer Chemistry* **34**, 8, pp. 1501–1509.

Chiamori, H. C., Brown, J. W., Adhiprakasha, E. V., Hantsoo, E. T., Straalsund, J. B., Melosh, N. A., and Pruitt, B. L. (2008). Suspension of nanoparticles in su-8: processing and characterization of nanocomposite polymers, *Microelectronics Journal* **39**, 2, pp. 228–236.

Chiang, W. H., Richmonds, C., and Sankaran, R. M. (2010). Continuous-flow, atmospheric-pressure microplasmas: a versatile source for metal nanoparticle synthesis in the gas or liquid phase, *Plasma Sources Science & Technology* **19**, 3, p. 034011.

Clark, T., Knox, T. M. O., McKervey, M. A., Mackle, H., and Rooney, J. J. (1979). Thermochemistry of bridged-ring substances: enthalpies of formation of some diamondoid hydrocarbons and of perhydroquinacene; comparisons with data from empirical force-field calculations, *Journal of the American Chemical Society* **101**, 9, pp. 2404–2410.

Clay, W. A., Liu, Z., Yang, W. L., Fabbri, J. D., Dahl, J. E., Carlson, R. M. K., Sun, Y., Schreiner, P. R., Fokin, A. A., Tkachenko, B. A., Fokina, N. A., Pianetta, P. A., Melosh, N., and Shen, Z. X. (2009). Origin of the monochromatic photoemission peak in diamondoid monolayers, *Nano Letters* **9**, 1, pp. 57–61.

Clay, W. A., Maldonado, J. R., Pianetta, P., Dahl, J. E. P., Carlson, R. M. K., Schreiner, P. R., Fokin, A. A., Tkachenko, B. A., Melosh, N. A., and Shen, Z. X. (2012). Photocathode device using diamondoid and cesium bromide films, *Applied Physics Letters* **101**, 24, p. 241605.

Clay, W. A., Sasagawa, T., Kelly, M., Dahl, J. E., Carlson, R. M. K., Melosh, N., and Shen, Z. X. (2008). Diamondoids as low-kappa dielectric materials, *Applied Physics Letters* **93**, 17, p. 172901.

Commenge, J. M., Falk, L., Corriou, J. P., and Matlosz, M. (2002). Optimal design for flow uniformity in microchannel reactors, *AIChE Journal* **48**, 2, pp. 345–358.

Courtney, T., Johnston, D. E., Rooney, J. J., and McKervey, M. A. (1972). Chemistry of diamantane. 1. synthesis and some functionalization reactions, *Journal of the Chemical Society: Perkin Transactions* **1**, 21, pp. 2691–2696.

Cui, J. B., Ristein, J., and Ley, L. (1998). Electron affinity of the bare and hydrogen covered single crystal diamond (111) surface, *Physical Review Letters* **81**, 2, pp. 429–432.

Cupas, C., Schleyer, P. V., and Trecker, D. J. (1965). Congressane, *Journal of the American Chemical Society* **87**, 4, pp. 917–+.

Curri, M. L., Comparelli, R., Striccoli, M., and Agostiano, A. (2010). Emerging methods for fabricating functional structures by patterning and assembling engineered nanocrystals, *Physical Chemistry Chemical Physics* **12**, 37, pp. 11197–11207.

Dahl, J. E., Liu, S. G., and Carlson, R. M. K. (2003a). Isolation and structure of higher diamondoids, nanometer-sized diamond molecules, *Science* **299**, 5603, pp. 96–99.

Dahl, J. E., Moldowan, J. M., Peters, K. E., Claypool, G. E., Rooney, M. A., Michael, G. E., Mello, M. R., and Kohnen, M. L. (1999). Diamondoid hydrocarbons as indicators of natural oil cracking, *Nature* **399**, 6731, pp. 54–57.

Dahl, J. E. P., Moldowan, J. M., Peakman, T. M., Clardy, J. C., Lobkovsky, E., Olmstead, M. M., May, P. W., Davis, T. J., Steeds, J. W., Peters, K. E., Pepper, A., Ekuan, A., and Carlson, R. M. K. (2003b). Isolation and structural proof of the large diamond molecule, cyclohexamantane (c26h30), *Angewandte Chemie-International Edition* **42**, 18, pp. 2040–2044.

Dahl, J. E. P., Moldowan, J. M., Wei, Z. B., Lipton, P. A., Denisevich, P., Gat, R., Liu, S. G., Schreiner, P. R., and Carlson, R. M. K. (2010). Synthesis of higher diamondoids and implications for their formation in petroleum, *Angewandte Chemie-International Edition* **49**, 51, pp. 9881–9885.

Dai, Z. R., Bradley, J. P., Joswiak, D. J., Brownlee, D. E., Hill, H. G. M., and Genge, M. J. (2002). Possible in situ formation of meteoritic nanodiamonds in the early solar system, *Nature* **418**, 6894, pp. 157–159.

Dang, T. D., Dalton, M. J., Venkatasubramanian, N., Johnson, J. A., Cerbus, C. A., and Feld, W. A. (2004). Synthesis and characterization of polyaryeneetherketone triphenylphosphine oxides incorporating cycloaliphatic/cage hydrocarbon structural units, *Journal of Polymer Science Part A-Polymer Chemistry* **42**, 23, pp. 6134–6142.

Daulton, T. L., Eisenhour, D. D., Bernatowicz, T. J., Lewis, R. S., and Buseck, P. R. (1996). Genesis of presolar diamonds: comparative high-resolution transmission electron microscopy study of meteoritic and terrestrial nano-diamonds, *Geochimica et Cosmochimica Acta* **60**, 23, pp. 4853–4872.

Davis, M. C., and Liu, S. G. (2006). Selective apical bromination of diamantane and conversion to the dihydroxy and dicarboxylic acid derivatives, *Synthetic Communications* **36**, 23, pp. 3509–3514.

Davis, M. C., and Nissan, D. A. (2006). Preparation of diamines of adamantane and diamantane from the diazides, *Synthetic Communications* **36**, 15, pp. 2113–2119.

DeCarli, P. S., and Jamieson, J. C. (1961). Formation of diamond by explosive shock, *Science* **133**, 346, pp. 1821–+.

Degen, C. L. (2008). Scanning magnetic field microscope with a diamond single-spin sensor, *Applied Physics Letters* **92**, 24, p. 243111.

del Alamo, J. A. (2011). Nanometre-scale electronics with iii-v compound semiconductors, *Nature* **479**, 7373, pp. 317–323.

Dheu, M. L., Gagnaire, D., Duddeck, H., Hollowood, F., and Mckervey, M. A. (1979). C-13 nuclear magnetic-resonance Spectra. 9. Strain effects on c-13 chemical-shifts of triamantane, *Journal of the Chemical Society-Perkin Transactions 2*, 3, pp. 357–359.

Dolmatov, V. Y., Myllymaki, V., and Vehanen, A. (2013). A possible mechanism of nanodiamond formation during detonation synthesis, *Journal of Superhard Materials* **35**, 3, pp. 143–150.

Dolmatov, V. Y., Veretennikova, M. V., Marchukov, V. A., and Sushchev, V. G. (2004). Currently available methods of industrial nanodiamond synthesis, *Physics of the Solid State* **46**, 4, pp. 611–615.

Drummond, N. D. (2007). Diamondoids display their potential, *Nature Nanotechnology* **2**, 8, pp. 462–463.

Drummond, N. D., Williamson, A. J., Needs, R. J., and Galli, G. (2005). Electron emission from diamondoids: a diffusion quantum monte carlo study, *Physical Review Letters* **95**, 9, p. 096801.

Duddeck, H., Hollowood, F., Karim, A., and McKervey, M. A. (1979). C-13 nuclear magnetic-resonance Spectra. 10. Substituent effects on the c-13 chemical-shifts of adamantanes, diamantanes, and triamantanes, *Journal of the Chemical Society-Perkin Transactions 2*, 3, pp. 360–365.

Dunk, P. W., Kaiser, N. K., Hendrickson, C. L., Quinn, J. P., Ewels, C. P., Nakanishi, Y., Sasaki, Y., Shinohara, H., Marshall, A. G., and Kroto, H. W. (2012). Closed network growth of fullerenes, *Nature Communications* **3**, p. 855.

Eckroth, D. R. (1967). A method for manual generation of correct von Baeyer names of polycyclic hydrocarbons, *Journal of Organic Chemistry* **32**, 11, pp. 3362–3365.

Edward, J. T., Farrell, P. G., and Shahidi, F. (1979). Partial molal volumes of organic-compounds in carbon-tetrachloride. 5. Cyclic alkanes, ethers, alcohols, ketones, and bromides, *Canadian Journal of Chemistry-Revue Canadienne De Chimie* **57**, 22, pp. 2892–2894.

Enoki, T., Takai, K., Osipov, V., Baidakova, M., and Vul, A. (2009). Nanographene and nanodiamond: new members in the nanocarbon family, *Chemistry: An Asian Journal* **4**, 6, pp. 796–804.

Eslamimanesh, A., Yazdizadeh, M., Mohammadi, A. H., and Richon, D. (2011). Experimental data assessment test for diamondoids solubility in gaseous system, *Journal of Chemical and Engineering Data* **56**, 5, pp. 2655–2659.

Evidente, V. G. H., Adler, C. H., Caviness, J. N., and Gwinn-Hardy, K. (1999). A pilot study on the motor effects of rimantadine in Parkinson's disease, *Clin. Neuropharmacol.* **22**, 1, pp. 30–32.

Fang, C. C., Xiong, Y. Q., Li, Y., Chen, Y., Liu, J. Z., Zhang, H. Z., Adedosu, T. A., and Peng, P. A. (2013). The origin and evolution of adamantanes and diamantanes in petroleum, *Geochimica et Cosmochimica Acta* **120**, pp. 109–120.

Fang, C. C., Xiong, Y. Q., Liang, Q. Y., and Li, Y. (2012). Variation in abundance and distribution of diamondoids during oil cracking, *Organic Geochemistry* **47**, pp. 1–8.

Farcasiu, D., Bohm, H., and Schleyer, P. V. (1977). Stepwise elaboration of diamondoid hydrocarbons: synthesis of diamantane from adamantane, *Journal of Organic Chemistry* **42**, 1, pp. 96–102.

Farooq, O., Farnia, S. M. F., Stephenson, M., and Olah, G. A. (1988). Superacid-catalyzed near-quantitative isomerization of c(4n+6)h(4n+12) (n=1−3) polycyclic precursors to diamondoid cage hydrocarbons promoted by 1-halodamantanes and sonication, *Journal of Organic Chemistry* **53**, 12, pp. 2840–2843.

Filik, J. (2010). *Diamondoid hydrocarbons, Carbon Based Nanomaterials*, Vol. 66–66, chap. 1 (Trans. Tech. Publications, Durnten-Zurich, Switzerland), pp. 1–26.

Filik, J., Harvey, J. N., Allan, N. L., May, P. W., Dahl, J. E. P., Liu, S., and Carlson, R. M. K. (2006a). Raman spectroscopy of nanocrystalline diamond: an ab initio approach, *Physical Review B* **74**, 3, p. 035423.

Filik, J., Harvey, J. N., Allan, N. L., May, P. W., Dahl, J. E. P., Liu, S. G., and Carlson, R. M. K. (2006b). Raman spectroscopy of diamondoids, *Spectrochimica Acta Part A: Molecular and Biomolecular Spectroscopy* **64**, 3, pp. 681–692.

Fokin, A. A., Gunchenko, P. A., Novikovsky, A. A., Shubina, T. E., Chernyaev, B. V., Dahl, J. E. P., Carlson, R. M. K., Yurchenko, A. G., and Schreiner, P. R. (2009). Photoacetylation of diamondoids: selectivities and mechanism, *European Journal of Organic Chemistry*, 30, pp. 5153–5161.

Fokin, A. A., and Schreiner, P. R. (2009). Band gap tuning in nanodiamonds: first principle computational studies, *Molecular Physics* **107**, 8–12, pp. 823–830.

Fokin, A. A., Schreiner, P. R., Fokina, N. A., Tkachenko, B. A., Hausmann, H., Serafin, M., Dahl, J. E. P., Liu, S. G., and Carlson, R. M. K. (2006). Reactivity of [1(2,3)4]pentamantane (t-d-pentamantane): a nanoscale model of diamond, *Journal of Organic Chemistry* **71**, 22, pp. 8532–8540.

Fokin, A. A., Tkachenko, B. A., Gunchenko, P. A., Gusev, D. V., and Schreiner, P. R. (2005). Functionalized nanodiamonds part I. An experimental assessment of diamantane and computational predictions for higher diamondoids, *Chemistry-A European Journal* **11**, 23, pp. 7091–7101.

Fokina, N. A., Tkachenko, B. A., Merz, A., Serafin, M., Dahl, J. E. P., Carlson, R. M. K., Fokin, A. A., and Schreiner, P. R. (2007). Hydroxy derivatives of diamantane, triamantane, and [121]tetramantane: selective preparation of bis-apical derivatives, *European Journal of Organic Chemistry*, 28, pp. 4738–4745.

Fort, R. C., and Schleyer, P. V. (1964). Adamantane: consequences of diamondoid structure, *Chemical Reviews* **64**, 3, pp. 277–300.

Freitas, R. A. (2009). Meeting the challenge of building diamondoid medical nanorobots, *International Journal of Robotics Research* **28**, 4, pp. 548–557.

Freitas, R. A., and Merkle, R. C. (2008). A minimal toolset for positional diamond mechanosynthesis, *Journal of Computational and Theoretical Nanoscience* **5**, 5, pp. 760–861.

Garcia, J. C., Assali, L. V. C., Machado, W. V. M., and Justo, J. F. (2010a). Crystal engineering using functionalized adamantane, *Journal of Physics-Condensed Matter* **22**, 31, p. 315303.

Garcia, J. C., Justo, J. F., Machado, W. V. M., and Assali, L. V. C. (2009). Functionalized adamantane: building blocks for nanostructure self-assembly, *Physical Review B* **80**, 12, p. 125421.

Garcia, J. C., Justo, J. F., Machado, W. V. M., and Assali, L. V. C. (2010b). Boron and nitrogen functionalized diamondoids: a first principles investigation, *Diamond and Related Materials* **19**, 7–9, pp. 837–840.

Garcia, J. C., Justo, J. F., Machado, W. V. M., and Assali, L. V. C. (2010c). Structural, electronic, and vibrational properties of amino-adamantane and rimantadine isomers, *Journal of Physical Chemistry A* **114**, 44, pp. 11977–11983.

Garcia, J. C., Machado, W. V. M., Assali, L. V. C., and Justo, J. F. (2011). Transition metal atoms encapsulated in adamantane molecules, *Diamond and Related Materials* **20**, 8, pp. 1222–1224.

Ghosh, A., Sciamanna, S. F., Dahl, J. E., Liu, S. G., Carlson, R. M. K., and Schiraldi, D. A. (2007). Effect of nanoscale diamondoids on the thermomechanical and morphological behaviors of polypropylene and polycarbonate, *Journal of Polymer Science Part B-Polymer Physics* **45**, 9, pp. 1077–1089.

Giruts, M. V., Rusinova, G. V., and Gordadze, G. N. (2006). Generation of adamantanes and diamantanes by thermal cracking of high-molecular-mass saturated fractions of crude oils of different genotypes, *Petroleum Chemistry* **46**, 4, pp. 225–236.

Giruts, M. V., and Gordadze, G. N. (2007). Generation of adamantanes and diamantanes by thermal cracking of polar components of crude oils of different genotypes, *Petroleum Chemistry* **47**, 1, pp. 12–22.

Greensfelder, B. S., Voge, H. H., and Good, G. M. (1949). Catalytic and thermal cracking of pure hydrocarbons: mechanisms of reaction, *Industrial and Engineering Chemistry* **41**, 11, pp. 2573–2584.

Gund, T. M., Nomura, M., and Schleyer, P. V. (1974a). Diamantane. 2. Preparation of derivatives of diamantane, *Journal of Organic Chemistry* **39**, 20, pp. 2987–2994.

Gund, T. M., Osawa, E., Williams, V. Z., and Schleyer, P. V. (1974b). Diamantane. 1. Preparation of diamantane: physical and spectral properties, *Journal of Organic Chemistry* **39**, 20, pp. 2979–2987.

Gund, T. M., Williams, V. Z., Osawa, E., and Schleyer, P. V. (1970). Convenient, high-yield preparation of diamantane (congressane) convenient, high-yield preparation of diamantane (congressane), *Tetrahedron Letters* **11**, 44, pp. 3877–3880.

Hajek, M., Vodicka, L., Trska, P., and Sklenar, V. (1985). C-13 NMR-spectra of adamantane carboxylic-acids and diamantane mono-carboxylic and di-carboxylic acids and esters, *Magnetic Resonance in Chemistry* **23**, 1, pp. 57–60.

Hala, S., Landa, S., and Hanus, V. (1966). Isolation of tetracyclo [6.3.1.02,6.0510]dodecane and pentacyclo[7.3.1.14,12.02,7.06,11] tetradecane (diamantane) from petroleum, *Angewandte Chemie, International Edition* **5**, 12, pp. 1045–1046.

Hamilton, R., McKervey, M. A., Rooney, J. J., and Malone, J. F. (1976). Gas-phase reactions on platinum: short synthesis of triamantane, *Journal of the Chemical Society, Chemical Communications*, 24, pp. 1027–1028.

Hamilton, R., Rooney, J. J., Johnston, D. E., and McKervey, M. A. (1972). Gas-phase conformational preference of methyl group from equilibration of 1-,3-, and 4-methyldiamantane on a chlorinated platinum-alumin catalyst, *Journal of the Chemical Society, Chemical Communications*, 22, pp. 1209–&.

Harris, S. J. (1990). Mechanism for diamond growth from methyl radicals, *Applied Physics Letters* **56**, 23, pp. 2298–2300.

Hartman, R. L., and Jensen, K. F. (2009). Microchemical systems for continuous-flow synthesis, *Lab on a Chip* **9**, 17, pp. 2495–2507.

Heurlin, M., Magnusson, M. H., Lindgren, D., Ek, M., Wallenberg, L. R., Deppert, K., and Samuelson, L. (2012). Continuous gas-phase synthesis of nanowires with tunable properties, *Nature* **492**, 7427, pp. 90–+.

Hill, H. G. M., Jones, A. P., and d'Hendecourt, L. B. (1998). Diamonds in carbon-rich proto-planetary nebulae, *Astronomy & Astrophysics* **336**, 3, pp. L41–L44.

Hodek, P., Janscak, P., Anzenbacher, P., Burkhard, J., Janku, J., and Vodicka, L. (1988). Metabolism of diamantane by rat-liver microsomal cytochromes-p-450, *Xenobiotica* **18**, 10, pp. 1109–1118.

Hohman, J. N., Claridge, S. A., Kim, M., and Weiss, P. S. (2010). Cage molecules for self-assembly, *Materials Science & Engineering R-Reports* **70**, 3–6, pp. 188–208.

Hollowood, F., Karim, A., McKervey, M. a., McSweeney, P., and Duddeck, H. (1978). Regioselective functionalization of triamantane, *Journal of the Chemical Society, Chemical Communications*, 7, pp. 306–308.

Hollowood, F. S., McKervey, M. A., Hamilton, R., and Rooney, J. J. (1980). Synthesis of triamantane, *Journal of Organic Chemistry* **45**, 24, pp. 4954–4958.

Hopf, H. (2003). Diamonds from crude oil? *Angewandte Chemie, International Edition* **42**, 18, pp. 2000–2002.

Horska, A., Podehradska, J., Vodicka, L., Hajek, M., and Trska, P. (1988). C-13 NMR-spectra of some disubstituted diamantanoid compounds, *Magnetic Resonance in Chemistry* **26**, 1, pp. 33–36.

Hu, Y. H., Shenderova, O. A., Hu, Z., Padgett, C. W., and Brenner, D. W. (2006). Carbon nanostructures for advanced composites, *Reports on Progress in Physics* **69**, 6, pp. 1847–1895.

Huang, L., Zhang, S. C., Wang, H. T., Fu, X. F., Zhang, W. L., Xu, Y. R., and Wei, C. Y. (2011). A novel method for isolation of diamondoids from crude oils for compound-specific isotope analysis, *Organic Geochemistry* **42**, 5, pp. 566–571.

Humphrey, W., Dalke, A., and Schulten, K. (1996). VMD: visual molecular dynamics, *Journal of Molecular Graphics* **14**, 1, pp. 33–38.

Ishii, C., Stauss, S., Kuribara, K., Urabe, K., Sasaki, T., and Terashima, K. (2015). Atmospheric pressure synthesis of diamondoids by plasmas generated inside a microfluidic reactor, *Diamond and Related Materials* **59**, pp. 40–46.

Iwai, H. (2009). Roadmap for 22 nm and beyond, *Microelectronic Engineering* **86**, 7–9, pp. 1520–1528.

Iwasa, A., Clay, W. A., Dahl, J. E., Carlson, R. M. K., Shen, Z. X., and Sasagawa, T. (2010). Environmentally friendly refining of diamond-molecules via the growth of large single crystals, *Crystal Growth & Design* **10**, 2, pp. 870–873.

Iza, F., Kim, G. J., Lee, S. M., Lee, J. K., Walsh, J. L., Zhang, Y. T., and Kong, M. G. (2008). Microplasmas: sources, particle kinetics, and biomedical applications, *Plasma Process. Polym.* **5**, 4, pp. 322–344.

Janku, J., Burkhard, J., and Vodicka, L. (1981). Hydrolysis of bromine derivatives of diamantane with nitric-acid, *Zeitschrift für Chemie* **21**, 9, pp. 325–326.

Jenkins, T. E. (1978). Low-frequency light-scattering in triamantane between 40-k and 340-k, *Journal of Physics C-Solid State Physics* **11**, 23, pp. L935–L937.

Jenkins, T. E., and Lewis, J. (1980). Raman-study of adamantane (c10h16), diamantane (c14h20) and triamantane (c18h24) between 10k and room temperatures, *Spectrochimica Acta Part A: Molecular and Biomolecular Spectroscopy* **36**, 3, pp. 259–264.

Jenkins, T. E., and O'Brien, P. (1981). A room-temperature specific-heat anomaly in triamantane (c18h24), *Physica Status Solidi A: Applied Research* **67**, 2, pp. K161–K162.

Jessop, P. G., and Leitner, W. (eds.) (1999). *Chemical Synthesis Using Supercritical Fluids* (Wiley-VCH, Verlag GmbH).

Jones, S. R., and Mellor, J. M. (1976). Bridgehead functionalization of saturated-hydrocarbons with lead(iv) salts, *Journal of the Chemical Society-Perkin Transactions* **1**, 23, pp. 2576–2581.

Karasek, P., Planeta, J., and Roth, M. (2008). Solubilities of adamantane and diamantane in pressurized hot water, *Journal of Chemical and Engineering Data* **53**, 3, pp. 816–819.

Karle, I. L., and Karle, J. (1965). Crystal and molecular structure of congressane c14h20 by x-ray diffraction, *Journal of the American Chemical Society* **87**, 4, pp. 918–920.

Karlen, S. D., Ortiz, R., Chapman, O. L., and Garcia-Garibay, M. A. (2005). Effects of rotational symmetry order on the solid state dynamics of phenylene and diamantane rotators, *Journal of the American Chemical Society* **127**, 18, pp. 6554–6555.

Kaviani, M., Deak, P., Aradi, B., Kohler, T., and Frauenheim, T. (2013). How small nanodiamonds can be? MD study of the stability against graphitization, *Diamond and Related Materials* **33**, pp. 78–84.

Kawashima, A., Toyota, H., Nomura, S., Takemori, T., Mukasa, S., Maehara, T., and Yamashita, H. (2007). 27.12 MHz plasma generation in supercritical carbon dioxide, *J. Appl. Phys.* **101**, 9, p. 093303.

Kikuchi, H., Kubo, H., Tomai, T., and Terashima, K. (2008). Cu films deposition by dielectric barrier discharge in supercritical CO_2, Ar and Xe environments, *Thin Solid Films* **516**, 19, pp. 6677–6682.

Kikuchi, H., Stauss, S., Nakahara, S., Matsubara, F., Tomai, T., Sasaki, T., and Terashima, K. (2010). Development of sheet-like dielectric barrier discharge microplasma generated in supercritical fluids and its application to the synthesis of carbon nanomaterials, *Journal of Supercritical Fluids* **55**, 1, pp. 325–332.

Kogelschatz, U. (2007). Applications of microplasmas and microreactor technology, *Contributions to Plasma Physics* **47**, 1–2, pp. 80–88.

Kraska, T., Leonhard, K. O., Tuma, D., and Schneider, G. M. (2002). Correlation of the solubility of low-volatile organic compounds in near- and supercritical fluids. Part 1: applications to adamantane and beta-carotene, *Journal of Supercritical Fluids* **23**, 3, pp. 209–224.

Krishnamurthy, V. V., Shih, J. G., and Olah, G. A. (1985). Substituent effects on C-13 NMR chemical-shifts and one-bond C-13-C-13 coupling-constants in 1-substituted and 4-substituted diamantanes, *Journal of Organic Chemistry* **50**, 8, pp. 1161–1164.

Krueger, A. (2008). New carbon materials: biological applications of functionalized nanodiamond materials, *Chemistry: A European Journal* **14**, 5, pp. 1382–1390.

Krueger, A. (2010). *Carbon Nanomaterials and Nanotechnology* (Wiley-VCH, Weinheim).

Krueger, A. (2011). Beyond the shine: recent progress in applications of nanodiamond, *Journal of Materials Chemistry* **21**, 34, pp. 12571–12578.

Krueger, A., and Lang, D. (2012). Functionality is key: recent progress in the surface modification of nanodiamond, *Advanced Functional Materials* **22**, 5, pp. 890–906.

Lacoste, D. A., Muneoka, H., Guiberti, T. F., Pai, D. Z., Urabe, K., Stauss, S., and Terashima, K. (2014). Effect of light irradiation on townsend breakdown in helium at 5.2k, *Plasma Sources Science & Technology* **23**, 1, p. 012002.

Landt, L., Staiger, M., Wolter, D., Klunder, K., Zimmermann, P., Willey, T. M., van Buuren, T., Brehmer, D., Schreiner, P. R., Tkachenko, B. A., Fokin, A. A., Moller, T., and Bostedt, C. (2010). The influence of a single thiol group on the electronic and optical properties of the smallest diamondoid adamantane, *Journal of Chemical Physics* **132**, 2, p. 024710.

Larionova, I., Kuznetsov, V., Frolov, A., Shenderova, O., Moseenkov, S., and Mazov, I. (2006). Properties of individual fractions of detonation nanodiamond, *Diamond and Related Materials* **15**, 11–12, pp. 1804–1808.

Leddy, B. P., McKervey, M. A., and McSweeney, P. (1980). Some comments on the thermal-stability of substituted ammonium, phosphonium, and arsonium permanganates and their use in alkane oxidation, *Tetrahedron Letters* **21**, 23, pp. 2261–2264.

Lengger, S. K., Scarlett, A. G., West, C. E., and Rowland, S. J. (2013). Diamondoid diacids (' o-4 ' species) in oil sands process-affected water, *Rapid Communications in Mass Spectrometry* **27**, 23, pp. 2648–2654.

Lenzke, K., Landt, L., Hoener, M., Thomas, H., Dahl, J. E., Liu, S. G., Carlson, R. M. K., Moller, T., and Bostedt, C. (2007). Experimental determination of the ionization potentials of the first five members of the nanodiamond series, *Journal of Chemical Physics* **127**, 8, p. 084320.

Lewis, R. S., Ming, T., Wacker, J. F., Anders, E., and Steel, E. (1987). Interstellar diamonds in meteorites, *Nature* **326**, 6109, pp. 160–162.

Liang, Q. Y., Xiong, Y. Q., Fang, C. C., and Li, Y. (2012). Quantitative analysis of diamondoids in crude oils using gas chromatography-triple quadrupole mass spectrometry, *Organic Geochemistry* **43**, pp. 83–91.

Licklider, L., Kuhr, W. G., Lacey, M. P., Keough, T., Purdon, M. P., and Takigiku, R. (1995). Online microreactors capillary electrophoresis mass-spectrometry for the analysis of proteins and peptides, *Analytical Chemistry* **67**, 22, pp. 4170–4177.

Lin, P. A., Kumar, A., and Sankaran, R. M. (2012). New insights into plasma-assisted dissociation of organometallic vapors for gas-phase synthesis of metal nanoparticles, *Plasma Processes and Polymers* **9**, 11–12, pp. 1184–1193.

Lin, R., and Wilk, Z. A. (1995). Natural occurrence of tetramantane (c22h28) pentamantane (c26h32) and hexamantane (c30h36) in a deep petroleum reservoir, *Fuel* **74**, 10, pp. 1512–1521.

Lipavska, H., Vodicka, L., Burda, J., Triska, J., Krumphanzl, V., Vanek, Z., and Podojil, M. (1982). The screening of microorganisms for hydroxylation activity - hydroxylation of diamantanols, *Biotechnology Letters* **4**, 9, pp. 563–566.

Lou, Z. S., Chen, Q. W., Wang, W., and Zhang, Y. F. (2003). Synthesis of carbon nanotubes by reduction of carbon dioxide with metallic lithium, *Carbon* **41**, 15, pp. 3063–3067.

Lu, A. J., Pan, B. C., and Han, J. G. (2005). Electronic and vibrational properties of diamondlike hydrocarbons, *Physical Review B* **72**, 3, p. 035447.

Maier, F., Ristein, J., and Ley, L. (2001). Electron affinity of plasma-hydrogenated and chemically oxidized diamond (100) surfaces, *Physical Review B* **64**, 16, pp. art. no.-165411.

Malik, A. A., Archibald, T. G., Baum, K., and Unroe, M. R. (1991). New high-temperature polymers based on diamantane, *Macromolecules* **24**, 19, pp. 5266–5268.

Mansuy, L., Philp, R. P., and Allen, J. (1997). Source identification of oil spills based on the isotopic composition of individual components in weathered oil samples, *Environmental Science & Technology* **31**, 12, pp. 3417–3425.

Marchand, A. P. (1995). Polycyclic cage compounds: reagents, substrates, and materials for the 21st century, *Aldrichimica Acta* **28**, 4, pp. 95–104.

Marcon, L., Riquet, F., Vicogne, D., Szunerits, S., Bodart, J. F., and Boukherroub, R. (2010). Cellular and in vivo toxicity of functionalized nanodiamond in xenopus embryos, *Journal of Materials Chemistry* **20**, 37, pp. 8064–8069.

Marre, S., Adamo, A., Basak, S., Aymonier, C., and Jensen, K. F. (2010). Design and packaging of microreactors for high pressure and high temperature applications, *Industrial & Engineering Chemistry Research* **49**, 22, pp. 11310–11320.

Marre, S., and Jensen, K. F. (2010). Synthesis of micro and nanostructures in microfluidic systems, *Chemical Society Reviews* **39**, 3, pp. 1183–1202.

Marre, S., Park, J., Rempel, J., Guan, J., Bawendi, M. G., and Jensen, K. F. (2008). Supercritical continuous-microflow synthesis of narrow size distribution quantum dots, *Advanced Materials* **20**, 24, pp. 4830–+.

Marsusi, F., and Mirabbaszadeh, K. (2009). Altering the electronic properties of diamondoids through encapsulating small particles, *Journal of Physics-Condensed Matter* **21**, 21, p. 215303.

Mason, B. P., Price, K. E., Steinbacher, J. L., Bogdan, A. R., and McQuade, D. T. (2007). Greener approaches to organic synthesis using microreactor technology, *Chemical Reviews* **107**, 6, pp. 2300–2318.

May, P. W., Harvey, J. N., Smith, J. A., and Mankelevich, Y. A. (2006). Reevaluation of the mechanism for ultrananocrystalline diamond deposition from Ar/CH_4/H-2 gas mixtures, *Journal of Applied Physics* **99**, 10, p. 104907.

McKervey, M. A. (1980). Synthetic approaches to large diamondoid hydrocarbons, *Tetrahedron* **36**, 8, pp. 971–992.

McMullen, J. P., and Jensen, K. F. (2010). Integrated microreactors for reaction automation: new approaches to reaction development, *Annual Review of Analytical Chemistry, Vol. 3* **3**, pp. 19–42.

Meador, M. A. (1998). Recent advances in the development of processable high-temperature polymers, *Annual Review of Materials Science* **28**, pp. 599–630.

Meinke, R., Richter, R., Moller, T., Tkachenko, B. A., Schreiner, P. R., Thomsen, C., and Maultzsch, J. (2013). Experimental and theoretical Raman analysis of functionalized diamantane, *Journal of Physics B: Atomic Molecular and Optical Physics* **46**, 2, p. 025101.

Mitra, T., Jelfs, K. E., Schmidtmann, M., Ahmed, A., Chong, S. Y., Adams, D. J., and Cooper, A. I. (2013). Molecular shape sorting using molecular organic cages, *Nature Chemistry* **5**, 4, pp. 276–281.

Mizuochi, N., Makino, T., Kato, H., Takeuchi, D., Ogura, M., Okushi, H., Nothaft, M., Neumann, P., Gali, A., Jelezko, F., Wrachtrup, J., and Yamasaki, S. (2012). Electrically driven single-photon source at room temperature in diamond, *Nature Photonics* **6**, 5, pp. 299–303.

Mochalin, V. N., Pentecost, A., Li, X. M., Neitzel, I., Nelson, M., Wei, C. Y., He, T., Guo, F., and Gogotsi, Y. (2013). Adsorption of drugs on nanodiamond: toward development of a drug delivery platform, *Molecular Pharmaceutics* **10**, 10, pp. 3728–3735.

Mochalin, V. N., Shenderova, O., Ho, D., and Gogotsi, Y. (2012). The properties and applications of nanodiamonds, *Nature Nanotechnology* **7**, 1, pp. 11–23.

Mohan, N., Chen, C. S., Hsieh, H. H., Wu, Y. C., and Chang, H. C. (2010). In vivo imaging and toxicity assessments of fluorescent nanodiamonds in caenorhabditis elegans, *Nano Letters* **10**, 9, pp. 3692–3699.

Moore, G. E. (1998). Cramming more components onto integrated circuits (reprinted from electronics, pp. 114–117, April 19, 1965), *Proceedings of the IEEE* **86**, 1, pp. 82–85.

Motiei, M., Hacohen, Y. R., Calderon-Moreno, J., and Gedanken, A. (2001). Preparing carbon nanotubes and nested fullerenes from supercritical CO_2 by a chemical reaction, *Journal of the American Chemical Society* **123**, 35, pp. 8624–8625.

Mrowca, J. J., and Katz, T. J. (1966). Catalysis of a cycloaddition reaction by rhodium on carbon, *Journal of the American Chemical Society* **88**, 17, pp. 4012–4015.

Nakahara, S., Stauss, S., Kato, T., Sasaki, T., and Terashima, K. (2011). Synthesis of higher diamondoids by pulsed laser ablation plasmas in supercritical co(2), *Journal of Applied Physics* **109**, 12, p. 123304.

Nakahara, S., Stauss, S., Miyazoe, H., Shizuno, T., Suzuki, M., Kataoka, H., Sasaki, T., and Terashima, K. (2010). Pulsed laser ablation synthesis of diamond molecules in supercritical fluids, *Applied Physics Express* **3**, 9, p. 096201.

Nakazaki, M., Naemura, K., Chikamatsu, H., Iwasaki, M., and Hashimoto, M. (1981). Synthesis and absolute-configuration of optically active-d3-tritwistane, the gyrochiral prototype of twist diamond, *Journal of Organic Chemistry* **46**, 11, pp. 2300–2306.

Nasir, S., and Fazeelat, T. (2013). Diamondoid hydrocarbons as maturity indicators for condensates from Southern Indus basin, Pakistan, *Journal of Chemistry* **2013**, p. 636845.

Neumann, R., Khenkin, A. M., and Dahan, M. (1995). Hydroxylation of alkanes with molecular-oxygen catalyzed by a new ruthenium-substituted polyoxometalate, [wznru(2)(iii)(oh)(h2o)(znw9o34)2](11-), *Angewandte Chemie: International Edition in English* **34**, 15, pp. 1587–1589.

NIST/EPA/NIH Mass Spectral Library (EI) (2012). NIST/EPA/NIH Mass Spectral Library (EI).

Nordman, C. E., and Schmitkons, D. L. (1965). Phase transition and crystal structures of adamantane, *Acta Crystallographica* **18**, pp. 764–767.

Ogale, S. B., Patil, P. P., Phase, D. M., Bhandarkar, Y. V., Kulkarni, S. K., Kulkarni, S., Ghaisas, S. V., and Kanetkar, S. M. (1987). Synthesis of metastable phases via pulsed-laser-induced reactive quenching at liquid-solid interfaces, *Physical Review B* **36**, 16, pp. 8237–8250.

Oguchi, H., Sakka, T., and Ogata, Y. H. (2007). Effects of pulse duration upon the plume formation by the laser ablation of cu in water, *Journal of Applied Physics* **102**, 2, p. 023306.

Olah, G. A., Wu, A. H., Farooq, O., and Prakash, G. K. S. (1989). Synthetic methods and reactions. 135. Single-step reductive isomerization of unsaturated polycyclics to c(4n+6)h(4n+12) diamondoid cage hydrocarbons with sodium-borohydride triflic acid, *Journal of Organic Chemistry* **54**, 6, pp. 1450–1451.

Oomens, J., Polfer, N., Pirali, O., Ueno, Y., Maboudian, R., May, P. W., Filik, J., Dahl, J. E., Liu, S. G., and Carlson, R. M. K. (2006). Infrared spectroscopic investigation of higher diamondoids, *Journal of Molecular Spectroscopy* **238**, 2, pp. 158–167.

Osawa, E., Furusaki, a., Hashiba, N., Matsumoto, T., Singh, V., Tahara, Y., Wiskott, E., Farcasiu, M., Iizuka, T., Tanaka, N., Kan, T., and Schleyer, P. V. (1980). Application of force-field calculations to organic chemistry. 2. Thermodynamic rearrangements of larger polycyclic-hydrocarbons derived from the 38.5 and 41.5-degrees-C melting dimers of cyclooctatetraene: crystal and molecular-structures of 5-bromoheptacyclo[8.6.0.02,8.03,13.04,11.05,9.012,16]hexadecane (5-bromo-(c2)-bisethanobisnordiamantane), 6,12-dibromohepta-cyclo[7.7.0.02, 6.03,15.04, 12.05,10.011,16]hexadecane,an d nonacy-clo[11.7.1.12,18.03,16.04,13.05,10.06,14.07,11.015,20]docosane (bas-tardane)no, *Journal of Organic Chemistry* **45**, 15, pp. 2985–2995.

Oshima, F., Stauss, S., Ishii, C., Pai, D. Z., and Terashima, K. (2012). Plasma microreactor in supercritical xenon and its application to diamondoid synthesis, *Journal of Physics D: Applied Physics* **45**, 40, p. 402003.

Oshima, F., Stauss, S., Inose, Y., and Terashima, K. (2014). Synthesis and investigation of reaction mechanisms of diamondoids produced using plasmas generated inside microcapillaries in supercritical xenon, *Japanese Journal of Applied Physics* **53**, 1, p. 010214.

Panich, A. M., Shames, A. I., Vieth, H. M., Osawa, E., Takahashi, M., and Vul', A. Y. (2006). Nuclear magnetic resonance study of ultrananocrystalline diamonds, *European Physical Journal B* **52**, 3, pp. 397–402.

Patel, G. M., Patel, G. C., Patel, R. B., Patel, J. K., and Patel, M. (2006). Nanorobot: a versatile tool in nanomedicine, *Journal of Drug Targeting* **14**, 2, pp. 63–67.

Patil, P. P., Phase, D. M., Kulkarni, S. a., Ghaisas, S. V., Kulkarni, S. K., Kanetkar, S. M., Ogale, S. B., and Bhide, V. G. (1987). Pulsed-laser induced reactive quenching at a liquid-solid interface: aqueous oxidation of iron, *Physical Review Letters* **58**, 3, pp. 238–241.

Pichierri, F. (2006). Theoretical study of sila-adamantane, *Chemical Physics Letters* **421**, 4–6, pp. 319–323.

Pichot, V., Risse, B., Schnell, F., Mory, J., and Spitzer, D. (2013). Understanding ultrafine nanodiamond formation using nanostructured explosives, *Scientific Reports* **3**, p. 2159.

Pirali, O., Vervloet, M., Dahl, J. E., Carlson, R. M. K., Tielens, A. G. G. M. and Oomens, J. (2007). Infrared spectroscopy of diamondoid molecules: new insights into the presence of nanodiamonds in the interstellar medium, *Astrophysical Journal* **661**, 2, pp. 919–925.

Pittman, C. E., Cohen, L. H., and Meyer, H. (1982). Transport properties of helium near the liquid vapor critical point. 1. Thermal-conductivity of He-3, *J. Low Temp. Phys.* **46**, 1–2, pp. 115–135.

Polfer, N., Sartakov, B. G., and Oomens, J. (2004). The infrared spectrum of the adamantyl cation, *Chemical Physics Letters* **400**, 1–3, pp. 201–205.

Prelog, V., and Seiwerth, R. (1941a). A new productive representation of adamantinesc, *Berichte der Deutschen Chemischen Gesellschaft* **74**, pp. 1769–1772.

Prelog, V., and Seiwerth, R. (1941b). On the synthesis of adamantan, *Berichte der Deutschen Chemischen Gesellschaft* **74**, pp. 1644–1648.

Raizer, Y. P. (1991). *Gas Discharge Physics* (Springer).

Ramachandran, G., and Manogaran, S. (2006). Vibrational spectra of adarnantanes x10h16 and diamantanes x14h20 (x = C, Si, Ge, Sn): a theoretical study, *Journal of Molecular Structure: Theochem* **766**, 2–3, pp. 125–135.

Ramachandran, G., and Manogaran, S. (2007). Vibrational spectra of triamantane x18h24, iso-tetramantane x22h28 and cyclohexamantane x26h30 (x = C, Si, Ge, Sn): a theoretical study, *Journal of Molecular Structure: Theochem* **816**, 1–3, pp. 31–41.

Rander, T., Staiger, M., Richter, R., Zimmermann, T., Landt, L., Wolter, D., Dahl, J. E., Carlson, R. M. K., Tkachenko, B. A., Fokina, N. A., Schreiner, P. R., Moller, T., and Bostedt, C. (2013). Electronic structure tuning of diamondoids through functionalization, *Journal of Chemical Physics* **138**, 2, p. 024310.

Raty, J. Y., and Galli, G. (2003). Ultradispersity of diamond at the nanoscale, *Nature Materials* **2**, 12, pp. 792–795.

Raty, J. Y., and Galli, G. (2005). First principle study of nanodiamond optical and electronic properties, *Computer Physics Communications* **169**, 1–3, pp. 14–19.

Raty, J. Y., Galli, G., Bostedt, C., van Buuren, T. W., and Terminello, L. J. (2003). Quantum confinement and fullerenelike surface reconstructions in nanodiamonds, *Physcial Review Letters* **90**, 3, p. 037401.

Reisberg, B., Doody, R., Stoffler, A., Schmitt, F., Ferris, S., and Mobius, H. J. (2003). Memantine in moderate-to-severe Alzheimer's disease, *New England Journal of Medicine* **348**, 14, pp. 1333–1341.

Reiser, J., McGregor, E., Jones, J., Enick, R., and Holder, G. (1996). Adamantane and diamantane; phase diagrams, solubilities, and rates of dissolution, *Fluid Phase Equilibria* **117**, 1–2, pp. 160–167.

Richardson, S. L., Baruah, T., Mehl, M. J., and Pederson, M. R. (2005). Theoretical confirmation of the experimental raman spectra of the lower-order diamondoid molecule: cyclohexamantane (c26h30), *Chemical Physics Letters* **403**, 1–3, pp. 83–88.

Richardson, S. L., Baruah, T., Mehl, M. J., and Pederson, M. R. (2006). Cyclohexamantane (c26h30): first-principles DFT study of a novel diamondoid molecule, *Diamond and Related Materials* **15**, 4–8, pp. 707–710.

Roberts, P. J., and Ferguson, G. (1977). Anti-tetramantane, a large diamondoid fragment, *Acta Crystallographica Section B: Structural Science* **33**, JUL15, pp. 2335–2337.

Roth, S., Leuenberger, D., Osterwalder, J., Dahl, J. E., Carlson, R. M. K., Tkachenko, B. A., Fokin, A. A., Schreiner, P. R., and Hengsberger, M. (2010). Negative-electron-affinity diamondoid monolayers as high-brilliance source for ultrashort electron pulses, *Chemical Physics Letters* **495**, 1–3, pp. 102–108.

Saito, K., Takatani, K., Sakka, T., and Ogata, Y. H. (2002). Observation of the light emitting region produced by pulsed laser irradiation to a solid-liquid interface, *Applied Surface Science* **197**, pp. 56–60.

Saito, K., Sakka, T., and Ogata, Y. H. (2003). Rotational spectra and temperature evaluation of c-2 molecules produced by pulsed laser irradiation to a graphite-water interface, *Journal of Applied Physics* **94**, 9, pp. 5530–5536.

Saitow, K., and Yamamura, T. (2009). Effective cooling generates efficient emission: blue, green, and red light-emitting Si nanocrystals, *J. Phys. Chem. C* **113**, 19, pp. 8465–8470.

Sakka, T., Saito, K., and Ogata, Y. H. (2002a). Emission spectra of the species ablated from a solid target submerged in liquid: vibrational temperature of C-2 molecules in water-confined geometry, *Applied Surface Science* **197**, pp. 246–250.

Sakka, T., Takatani, K., Ogata, Y. H., and Mabuchi, M. (2002b). Laser ablation at the solid-liquid interface: transient absorption of continuous spectral emission by ablated aluminium atoms, *Journal of Physics D. Applied Physics* **35**, 1, pp. 65–73.

Sankaran, R. M., and Giapis, K. P. (2002). Hollow cathode sustained plasma microjets: characterization and application to diamond deposition, *Journal of Applied Physics* **92**, 5, pp. 2406–2411.

Sasagawa, T., and Shen, Z. X. (2008). A route to tunable direct band-gap diamond devices: electronic structures of nanodiamond crystals, *Journal of Applied Physics* **104**, 7, p. 073704.

Sasaki, K., and Aoki, S. (2008). Temporal variation of two-dimensional temperature in a laser-ablation plume produced from a graphite target, *Applied Physics Express* **1**, 8, p. 086001.

Savage, P. E., Gopalan, S., Mizan, T. I., Martino, C. J., and Brock, E. E. (1995). Reactions at supercritical conditions: applications and fundamentals, *AICHE J.* **41**, 7, pp. 1723–1778.

Schleyer, P. V. (1957). A simple preparation of adamantane, *Journal of the American Chemical Society* **79**, 12, pp. 3292–3292.

Schleyer, P. V. (1965). Synthesis of pentacyclo[7.3.1.14,12.02,7.06, 11]tetradecane, *Angewandte Chemie-International Edition* **4**, 2, pp. 169–170.

Schoell, M., and Carlson, R. M. K. (1999). Diamondoids and oil are not forever, *Nature* **399**, 6731, pp. 15–16.

Schrand, A. M., Dai, L., Schlager, J. J., Hussain, S. M., and Osawa, E. (2007a). Differential biocompatibility of carbon nanotubes and nanodiamonds, *Diamond and Related Materials* **16**, 12, pp. 2118–2123.

Schrand, A. M., Hens, S. A. C., and Shenderova, O. A. (2009). Nanodiamond particles: properties and perspectives for bioapplications, *Critical Reviews in Solid State and Materials Sciences* **34**, 1–2, pp. 18–74.

Schrand, A. M., Huang, H. J., Carlson, C., Schlager, J. J., Osawa, E., Hussain, S. M., and Dai, L. M. (2007b). Are diamond nanoparticles cytotoxic? *Journal of Physical Chemistry B* **111**, 1, pp. 2–7.

Schreiner, P. R., Fokina, N. A., Tkachenko, B. A., Hausmann, H., Serafin, M., Dahl, J. E. P., Liu, S. G., Carlson, R. M. K., and Fokin, A. A. (2006). Functionalized nanodiamonds: triamantane and [121]tetramantane, *Journal of Organic Chemistry* **71**, 18, pp. 6709–6720.

Schwertfeger, H., Fokin, A. A., and Schreiner, P. R. (2008). Diamonds are a chemist's best friend: diamondoid chemistry beyond adamantane, *Angewandte Chemie-International Edition* **47**, 6, pp. 1022–1036.

Schwertfeger, H., and Schreiner, P. R. (2010). Future of diamondoids, *Chemie in Unserer Zeit* **44**, 4, pp. 248–253.

Shen, M. Z., Schaefer, H. F., Liang, C. X., Lii, J. H., Allinger, N. L., and Schleyer, P. V. (1992). Finite td symmetry models for diamond: from adamantane to superadamantane (c35h36), *Journal of the American Chemical Society* **114**, 2, pp. 497–505.

Shenderova, O. A., Zhirnov, V. V., and Brenner, D. W. (2002). Carbon nanostructures, *Critical Reviews in Solid State and Material Sciences* **27**, 3–4, pp. 227–356.

Shimoyama, A., and Yabuta, H. (2002). Mono- and bicyclic alkanes and diamondoid hydrocarbons in the cretaceous/tertiary boundary sediments at Kawaruppu, Hokkaido, Japan, *Geochemical Journal* **36**, 2, pp. 173–189.

Shizuno, T., Miyazoe, H., Saito, K., Stauss, S., Suzuki, M., Sasaki, T., and Terashima, K. (2011). Synthesis of diamondoids by supercritical xenon discharge plasma, *Japanese Journal of Applied Physics* **50**, 3, p. 030207.

Sivaraman, G., and Fyta, M. (2014). Chemically modified diamondoids as biosensors for dna, *Nanoscale* **6**, 8, pp. 4225–4232.

Smith, V. S., and Teja, A. S. (1996). Solubilities of diamondoids in supercritical solvents, *Journal of Chemical and Engineering Data* **41**, 4, pp. 923–925.

Stauss, S., Ishii, C., Pai, D. Z., Urabe, K., and Terashima, K. (2014). Diamondoid synthesis in atmospheric pressure adamantane-argon-methane-hydrogen mixtures using a continuous flow plasma microreactor, *Plasma Sources Science & Technology* **23**, 3, p. 35016.

Stauss, S., Miyazoe, H., Shizuno, T., Saito, K., Sasaki, T., and Terashima, K. (2010). Synthesis of the higher-order diamondoid hexamantane using low-temperature plasmas generated in supercritical xenon, *Japanese Journal of Applied Physics* **49**, 7, p. 070213.

Stauss, S., Muneoka, H., Ebato, N., Oshima, F., Pai, D. Z., and Terashima, K. (2013). Self-organized pattern formation in helium dielectric barrier discharge cryoplasmas, *Plasma Sources Science & Technology* **22**, 2, p. 025021.

Steele, W. V., and Watt, I. (1977). Standard enthalpies of formation of adamantanoid compounds. 4. Methyladamantanes, *Journal of Chemical Thermodynamics* **9**, 9, pp. 843–849.

Steglich, M., Huisken, F., Dahl, J. E., Carlson, R. M. K., and Henning, T. (2011). Electronic spectroscopy of FUV-irradiated diamondoids: a combined experimental and theoretical study, *Astrophysical Journal* **729**, 2, p. 91.

Stout, S. A., and Douglas, G. S. (2004). Diamondoid hydrocarbons: application in the chemical fingerprinting of natural gas condensate and gasoline, *Environmental Forensics* **5**, 4, pp. 225–235.

Suga, Y., Tsuyoshi, N., and Watanabe, T. (2010). Generation of dielectric barrier discharge plasma in supercritical argon for material synthesis, *Japanese Journal of Applied Physics* **49**, p. 076002.

Swaid, I., Nickel, D., and Schneider, G. M. (1985). Nir-spectroscopic investigations on phase-behavior of low-volatile organic-substances in supercritical carbon-dioxide, *Fluid Phase Equilibria* **21**, 1–2, pp. 95–112.

Sweeney, J. J., and Burnham, A. K. (1990). Evaluation of a simple-model of vitrinite reflectance based on chemical-kinetics, *AAPG Bulletin-American Association of Petroleum Geologists* **74**, 10, pp. 1559–1570.

Tabushi, I., Kojo, S., Schleyer, P. V., and Gund, T. M. (1974). Selective functionalization of unactivated methine positions - 4-acetyldiamantane, *Journal of the Chemical Society, Chemical Communications*, 15, pp. 591–591.

Taylor, J. M., Cappellaro, P., Childress, L., Jiang, L., Budker, D., Hemmer, P. R., Yacoby, A., Walsworth, R., and Lukin, M. D. (2008). High-sensitivity diamond magnetometer with nanoscale resolution, *Nature Physics* **4**, 10, pp. 810–816.

Tiggelaar, R. M., Berenschot, J. W., de Boer, J. H., Sanders, R. G. P., Gardeniers, J. G. E., Oosterbroek, R. E., van den Berg, A., and Elwenspoek, M. C. (2005). Fabrication and characterization of high-temperature microreactors with thin film heater and sensor patterns in silicon nitride tubes, *Lab on a Chip* **5**, 3, pp. 326–336.

Tiwari, R. N., Tiwari, J. N., and Chang, L. (2010). The synthesis of diamond films on adamantane-coated Si substrate at low temperature, *Chemical Engineering Journal* **158**, 3, pp. 641–645.

Tiwari, R. N., Tiwari, J. N., Chang, L., and Yoshimura, M. (2011). Enhanced nucleation and growth of diamond film on Si by CVD using a chemical precursor, *Journal of Physical Chemistry C* **115**, 32, pp. 16063–16073.

Tkachenko, B. A., Fokina, N. A., Chernish, L. V., Dahl, J. E. P., Liu, S. G., Carlson, R. M. K., Fokin, A. A., and Schreiner, P. R. (2006). Functionalized nanodiamonds part 3: thiolation of tertiary/bridgehead alcohols, *Organic Letters* **8**, 9, pp. 1767–1770.

Tomai, T., Ito, T., and Terashima, K. (2006). Generation of dielectric barrier discharge in high-pressure N_2 and CO_2 environments up to supercritical conditions, *Thin Solid Films* **506**, pp. 409–413.

Tomai, T., Katahira, K., Kubo, H., Shimizu, Y., Sasaki, T., Koshizaki, N., and Terashima, K. (2007). Carbon materials syntheses using dielectric barrier discharge microplasma in supercritical carbon dioxide environments, *Journal of Supercritical Fluids* **41**, 3, pp. 404–411.

Trachsel, F., Tidona, B., Desportes, S., and von Rohr, P. R. (2009). Solid catalyzed hydrogenation in a si/glass microreactor using supercritical co2 as the reaction solvent, *Journal of Supercritical Fluids* **48**, 2, pp. 146–153.

Tsuda, M., Nakajima, M., and Oikawa, S. (1986). Epitaxial-growth mechanism of diamond crystal in CH_4-H_2 plasma, *Journal of the American Chemical Society* **108**, 19, pp. 5780–5783.

Tsuzuki, N., Hama, T., Kawada, M., Hasui, A., Konishi, R., Shiwa, S., Ochi, Y., Futaki, S., and Kitagawa, K. (1994). Adamantane as a brain-directed drug carrier for poorly absorbed drug. 2. AZT derivatives conjugated with the 1-adamantane moiety, *Journal of Pharmaceutical Sciences* **83**, 4, pp. 481–484.

Urabe, K., Kato, T., Stauss, S., Himeno, S., Kato, S., Muneoka, H., Baba, M., Suemoto, T., and Terashima, K. (2013). Dynamics of pulsed laser ablation in high-density carbon dioxide including supercritical fluid state, *Journal of Applied Physics* **114**, 14, p. 143303.

Vandenbroucke, M. (2003). Kerogen: from types to models of chemical structure, *Oil & Gas Science and Technology-Revue D IFP Energies Nouvelles* **58**, 2, pp. 243–269.

Verboom, W. (2009). Selected examples of high-pressure reactions in glass microreactors, *Chemical Engineering & Technology* **32**, 11, pp. 1695–1701.

Vesovic, V., Wakeham, W. A., Olchowy, G. A., Sengers, J. V., Watson, J. T. R., and Millat, J. (1990). The transport-properties of carbon-dioxide, *Journal of Physical and Chemical Reference Data* **19**, 3, pp. 763–808.

Voskuhl, J., Waller, M., Bandaru, S., Tkachenko, B. A., Fregonese, C., Wibbeling, B., Schreiner, P. R., and Ravoo, B. J. (2012). Nanodiamonds in sugar rings: an experimental and theoretical investigation of cyclodextrin-nanodiamond inclusion complexes, *Organic & Biomolecular Chemistry* **10**, 23, pp. 4524–4530.

Wada, N., Gaczi, P. J., and Solin, S. A. (1980). Diamond-like 3-fold coordinated amorphous-carbon, *Journal of Non-Crystalline Solids* **35-6**, JAN-, pp. 543–548.

Walas, S. M. (1985). *Phase Equilibria in Chemical Engineering* (Butterworth-Heinemann).

Wang, J. E., Chern, Y. T., and Chung, M. A. (1996). Synthesis and characterization of new poly(n-1-adamantylmaleimide) and poly(n-1-diamantylmaleimide), *Journal of Polymer Science Part A: Polymer Chemistry* **34**, 16, pp. 3345–3354.

Wang, J. J., Chang, Y. F., Chern, Y. T., and Chi, C. W. (2003). Study of in vitro and in vivo effects of 1,6-bis[4-(4-amino-3-hydroxy-phenoxy)phenyl]diamantane (dpd), a novel cytostatic and differentiation inducing agent, on human colon cancer cells, *British Journal of Cancer* **89**, 10, pp. 1995–2003.

Wei, Z. B., Moldowan, J. M., and Paytan, A. (2006). Diamondoids and molecular biomarkers generated from modern sediments in the absence and presence of minerals during hydrous pyrolysis, *Organic Geochemistry* **37**, 8, pp. 891–911.

Willey, T. M., Bostedt, C., van Buuren, T., Dahl, J. E., Liu, S. G., Carlson, R. M. K., Terminello, L. J., and Moller, T. (2005). Molecular limits to the quantum confinement model in diamond clusters, *Physical Review Letters* **95**, 11, p. 113401.

Willey, T. M., Fabbri, J. D., Lee, J. R. I., Schreiner, P. R., Fokin, A. A., Tkachenko, B. A., Fokina, N. A., Dahl, J. E. P., Carlson, R. M. K., Vance, A. L., Yang, W. L., Terminello, L. J., van Buuren, T., and Melosh, N. A. (2008). Near-edge x-ray absorption fine structure spectroscopy of diamondoid thiol monolayers on gold, *Journal of the American Chemical Society* **130**, 32, pp. 10536–10544.

Williams, O. A., Daenen, M., D'Haen, J., Haenen, K., Maes, J., Moshchalkov, V. V., Nesladek, M., and Gruen, D. M. (2006). Comparison of the growth and properties of ultrananocrystalline diamond and nanocrystalline diamond, *Diamond and Related Materials* **15**, 4–8, pp. 654–658.

Williams, V. Z., P.V., S., Gleicher, G. J., and Rodewald, L. B. (1966). Triamantane, *Journal of the American Chemical Society* **88**, 16, pp. 3862–3863.

Wingert, W. S. (1992). GC-MS analysis of diamondoid hydrocarbons in smackover petroleums, *Fuel* **71**, 1, pp. 37–43.

Xu, X. J. (2001). Dielectric barrier discharge: properties and applications, *Thin Solid Films* **390**, 1–2, pp. 237–242.

Yang, F., Lin, Y., Dahl, J. E. P., Carlson, R. M. K., and Mao, W. L. (2014). High pressure raman and x-ray diffraction study of [121] tetramantane, *Journal of Physical Chemistry C* **118**, 14, pp. 7683–7689.

Yang, W. L., Fabbri, J. D., Willey, T. M., Lee, J. R. I., Dahl, J. E., Carlson, R. M. K., Schreiner, P. R., Fokin, A. A., Tkachenko, B. A., Fokina, N. A., Meevasana, W., Mannella, N., Tanaka, K., Zhou, X. J., van Buuren, T., Kelly, M. A., Hussain, Z., Melosh, N. A., and Shen, Z. X. (2007). Monochromatic electron photoemission from diamondoid monolayers, *Science* **316**, 5830, pp. 1460–1462.

Yoshida, J. I., Nagaki, A., and Yamada, T. (2008). Flash chemistry: fast chemical synthesis by using microreactors, *Chemistry: A European Journal* **14**, 25, pp. 7450–7459.

Yuan, Y., Chen, Y. W., Lui, J. H., Wang, H. F., and Liu, Y. F. (2009). Biodistribution and fate of nanodiamonds in vivo, *Diamond and Related Materials* **18**, 1, pp. 95–100.

Yuan, Y., Wang, X., Jia, G., Liu, J. H., Wang, T. C., Gu, Y. Q., Yang, S. T., Zhen, S., Wang, H. F., and Liu, Y. F. (2010). Pulmonary toxicity and translocation of nanodiamonds in mice, *Diamond and Related Materials* **19**, 4, pp. 291–299.

Zhang, X. Y., Yin, J. L., Kang, C., Li, J., Zhu, Y., Li, W. X., Huang, Q., and Zhu, Z. Y. (2010). Biodistribution and toxicity of nanodiamonds in mice after intratracheal instillation, *Toxicology Letters* **198**, 2, pp. 237–243.

Index